P9-EDU-030

INTRODUCTION TO ELECTRICITY

Kurt Harding Schick
H. B. Beal Secondary School
London, Ontario

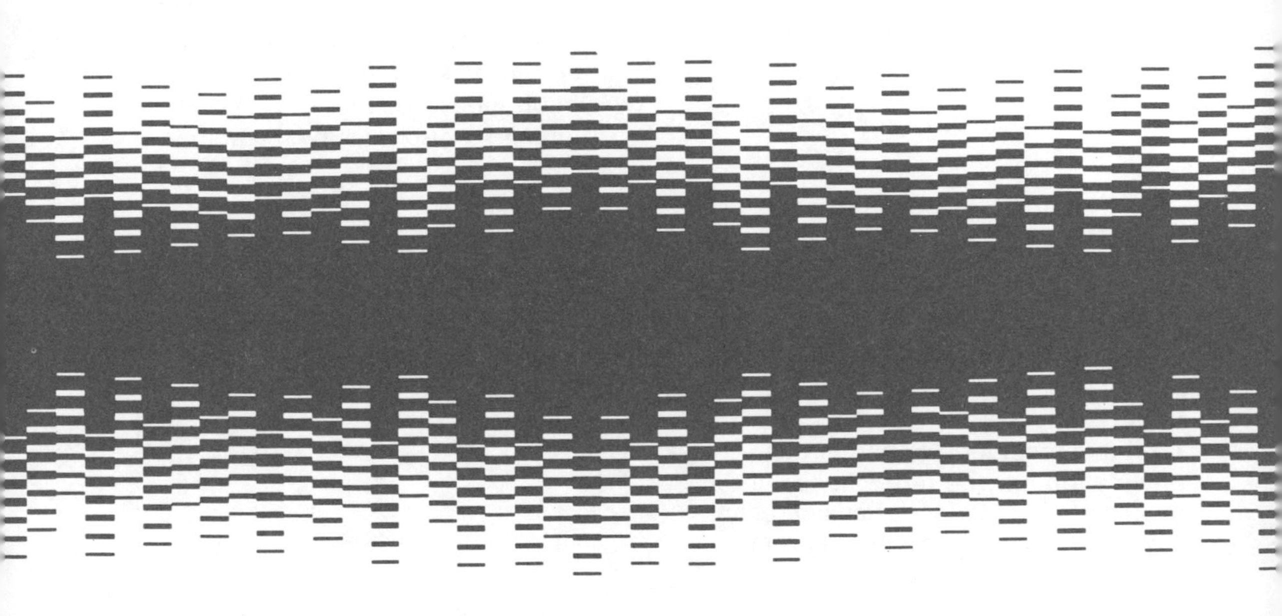

McGRAW-HILL RYERSON LIMITED

Toronto Montreal New York London Sydney
Johannesburg Mexico Panama Düsseldorf Singapore
São Paulo Kuala Lumpur New Delhi

I want to thank my fellow teacher Bob Lucas of the Beal Electrical Department for his careful work in checking this book for general accuracy and validity.

My thanks go also to Bill Green and Geoff Shilleto for their unfailing cooperation during the many months I spent in writing this book.

To Herb Hilderley of McGraw-Hill Ryerson I owe a special measure of gratitude for his valuable assistance and expert editorial guidance.

K.H.S.

Photos p. 1 and p. 51 (Figure 5-23) courtesy of Westinghouse Canada Ltd.
Photo p. 119 courtesy of Ontario Hydro.
Photo p. 124 (Figure 15-5) courtesy of Chubb Industries Ltd.
Remainder of photos by the author.

Introduction to Electricity

ISBN 0-07-077678-4

1 2 3 4 5 6 7 8 9 AP 4 3 2 1 0 9 8 7 6 5

Printed and bound in Canada

Table of Contents

PART ONE • FUNDAMENTALS OF ELECTRICITY

Chapter 10

Chapter 11

Chapter 12

Chapter 13

Chapter 14

PART TWO • APPLIED ELECTRICITY

Chapter 15

Chapter 16

Chapter 17

PART ONE
FUNDAMENTALS OF ELECTRICITY

1

ELECTRICITY: PULSE OF THE 20th CENTURY

You cannot appreciate the importance of electricity to our way of life until you have visited one of the pioneer villages to see how our forebears struggled along without it. Their machines and devices often were ingenious and many could still be used today. But they all lacked a common element: a built-in source of motive power. Saw mills would screech to life with a team of horses slowly pulling a huge horizontal wheel, while shafts and belts and cog-wheels transferred their power to the machine. Grinding wheels, washing machines

and other simple contrivances rumbled sluggishly, powered by a panting dog in a treadmill. Wooden water wheels slowly ground millstone against millstone. It was a hard and laborious existence. The day was never long enough to supply the common needs, and men and women laboured in the dark hours of the morning and far into the night by the flickering light of oil lamps or candles.

What a difference today! Gigantic generators humming with a low and steady monotone in huge power plants convert the energy of burning coal or falling water or atomic reactors to torrents of electric power. Overhead cables transport these vast amounts of energy to distant cities, towns and villages, even to individual farms. In factories and schools and hospitals, millions of electrical devices work silently and efficiently in the service of man. Electricity is at work everywhere, even in the distant outposts of civilization: polar camps, orbital space stations, and planetary probes.

In the wake of this electrification of society has come a tremendous expansion of job opportunities. Just as electricity has become indispensable to our way of life, so has risen the importance of the men and women who have made electricity their life's work. Literally hundreds of different careers are now based on a thorough knowledge of electricity, and to be competent in one of the many fields of electrical or electronic technology has become a source of great pride for many successful people.

Our way of life would be impossible without this silent, invisible, and clean form of energy. Electricity throbs through millions of cable arteries to billions of places to work for us. It has become the sustaining force of our modern society, the true pulse of the 20th century.

2

MAGNETS AND MAGNETISM

A magnet is a piece of iron oxide or special alloy that exerts an invisible force of attraction on objects made of iron, nickel, or cobalt.

The invisible force itself is called **magnetism** or **magnetic force.** When magnetism is doing work, it is referred to as **magnetic energy.**

Magnets are important to electricity. Many electrical devices depend on magnetic energy for their operation.

2-1 TYPES OF MAGNETS

Magnets fall into three distinct groups: natural magnets, artificial magnets, and electromagnets.

Natural magnets have been known for thousands of years. They were first found in the ancient Greek province of Magnesia and were therefore named Stones of Magnesia. All modern magnetic terms come from this old Greek name. The common English name for a natural magnet is **lodestone.**

Figure 2-1 Lodestones

Lodestones are pieces of blackish iron ore (magnetite) whose feeble magnetic force varies greatly from stone to stone.

Artificial magnets, or simply **permanent magnets**, are made of hard and brittle alloys composed of iron, nickel, cobalt, and other metals. These alloys are strongly magnetized during the manufacturing process. Permanent magnets come in many shapes and sizes; they serve many purposes in electrical equipment.

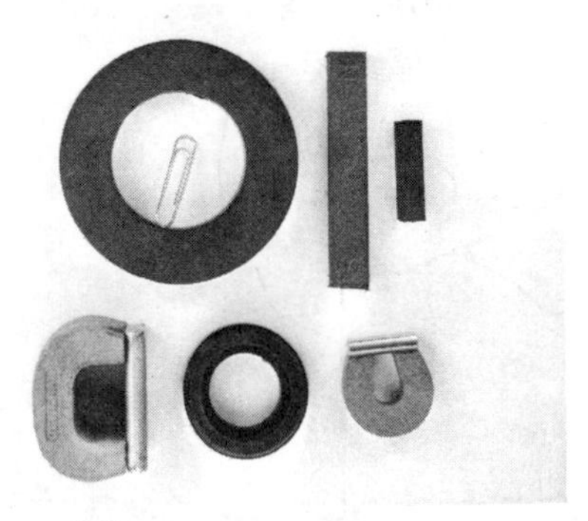

Figure 2-2 Permanent Magnets

Electromagnets are energy converters. They change electricity to magnetism. All electromagnets consist of two main parts: a core of special steel and a copper-wire coil wound on this core.

Lodestones and artificial magnets retain their magnetism indefinitely if handled properly. Electromagnets, on the other hand, can be turned on or off at will, and their magnetic force can be completely controlled.

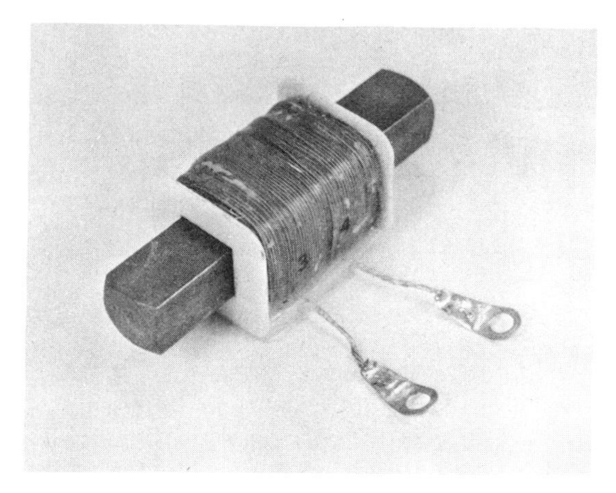

Figure 2-3 Electromagnet

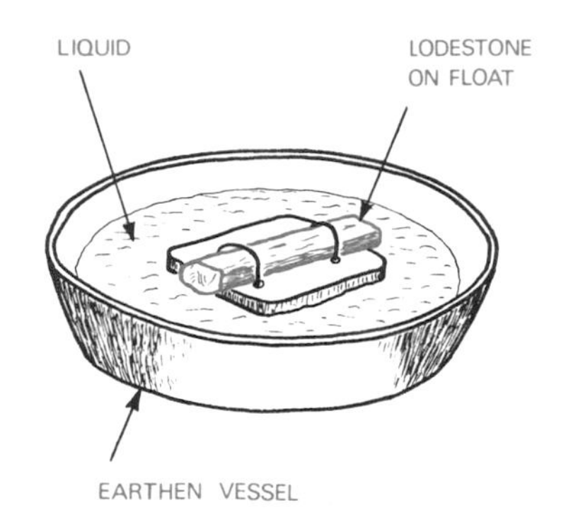

Figure 2-5 Medieval Lodestone Compass

2-2 THE MAGNETISM OF THE EARTH

Some eight hundred years ago, an unknown experimenter discovered a strange property of natural magnets. He found that a bar-shaped lodestone when suspended from a string would always **point toward the north star.**

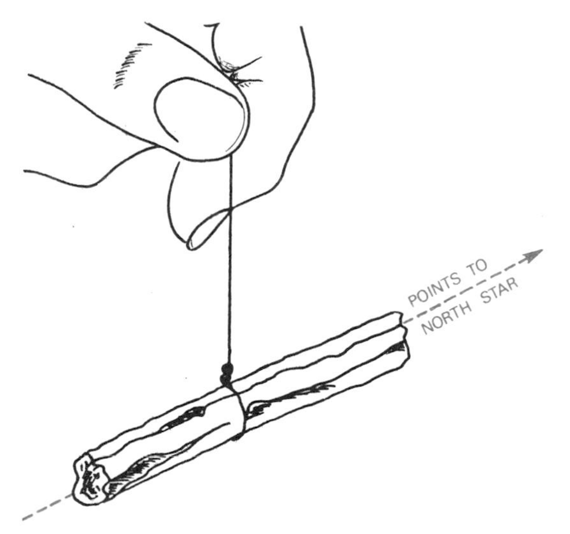

Figure 2-4 Lodestone Suspended from String

Beginning in the twelfth century, crude lodestone compasses were used to guide ships at sea; but no one knew why they worked.

In the year 1600, the English physician **William Gilbert** laid the foundations of our knowledge of magnetism. He published the results of eighteen years of experimentation with magnetism in a book entitled *De Magnete.* His chief discovery was that **the earth is a huge magnet** whose magnetic force is concentrated at magnetic poles located near the geographic poles. Gilbert found that it is the invisible magnetic force of these poles that turns the needle of a compass in a south-north direction.

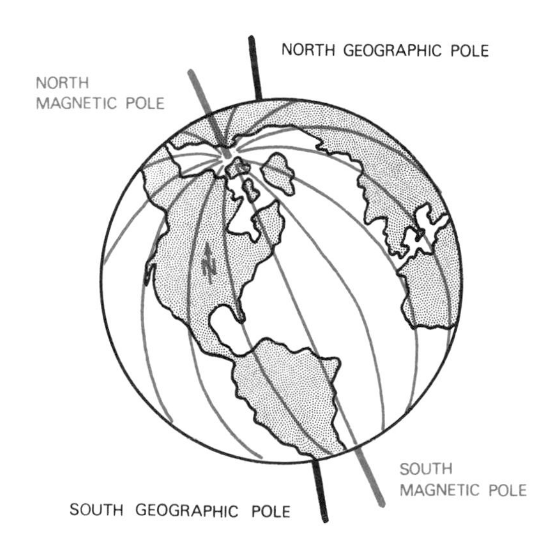

Figure 2-6 The Earth as a Magnet

Today we know the location of the magnetic poles quite accurately. The **magnetic north pole** lies some 1,100 miles (1,760 kilometres) south of the geographic north pole in northern Canada. The **magnetic south pole** is about 1,500 miles (2,400 kilometres)* from the geographic south pole on the Antarctic continent.

Both magnetic poles shift their location from year to year for reasons which are not yet fully understood.

*Metric calculations are approximate.

2-3 THE MODERN MAGNETIC COMPASS

The heart of the magnetic compass is its needle, a perfectly balanced permanent magnet. This needle magnet is pivoted on a sharply pointed metal post. The **north-seeking pole** of the compass needle is coloured for quick identification. The compass card is a white paper disc on which the four principal directions—north, east, south, and west—are marked. The paper disc is divided into 360 equal parts or degrees. With these calibrations, a direction can be stated in both words and degrees: 90° or due east, 135° or south-east, 180° or due south, and so on.

Figure 2-7 Modern Magnetic Compass

How to use a magnetic compass: place the compass on a flat surface or hold it steady in your hand, well away from steel objects. Wait until the needle has stopped swinging. Turn the compass slowly until the northseeking end of its needle lines up with the north mark on the compass card. The compass is now "oriented," with its marks pointing due north, south, east, and west.

A magnetic compass always deviates slightly from true north. The exact value of this difference can be obtained from the nearest weather station.

2-4 THE POLES OF A MAGNET

The location of the magnetic poles can be examined by dipping a bar magnet into a small pile of iron filings. The filings cling only to the ends of the magnet, where its magnetic force is concentrated.

Figure 2-8 Bar Magnet Dipped Into Iron Filings

A close look at individual filings shows that they seem to "point" toward a spot near each end of the magnet. These spots are the **magnetic poles**.

Figure 2-9 Magnetic Pole with Filings

A bar magnet suspended from a string always points the same end northward. This end is called its **northseeking pole**, or

Figure 2-10 Suspended Magnet

N-pole for short. The N-pole of a bar magnet is often coloured or stamped with a capital N.

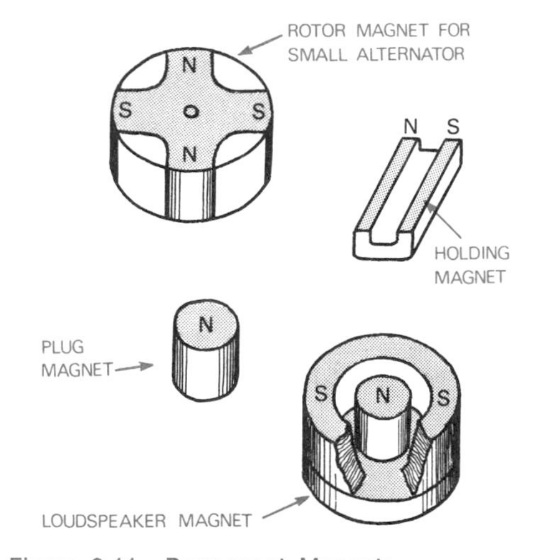

Figure 2-11 Permanent Magnets

2-5 LINES AND FIELDS OF MAGNETIC FORCE

If a bar magnet is covered with a sheet of paper, and if small iron filings are sprinkled over it, the magnetic force will align the filings into a definite pattern, as shown in Figure 2-12.

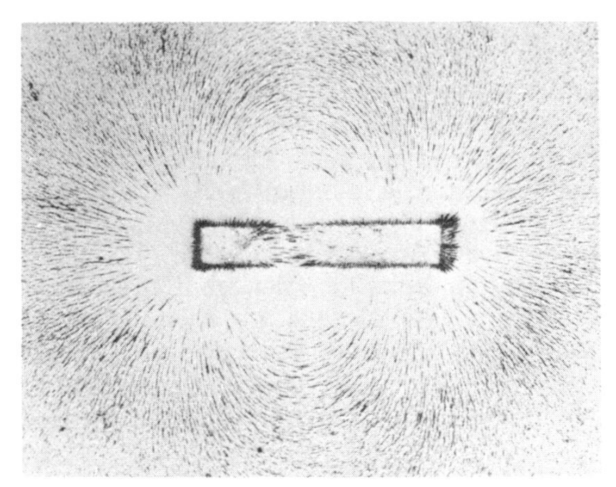

Figure 2-12 Iron-Filings Pattern of a Bar Magnet

This pattern shows that the magnetic force acts along thin, invisible paths that stretch from pole to pole in the space around the magnet. These paths are called **lines of magnetic force**, or **flux lines**. A closer inspection shows that many filings point upward and away from the magnetic poles, indicating that similar lines of force also exist above and below the magnet.

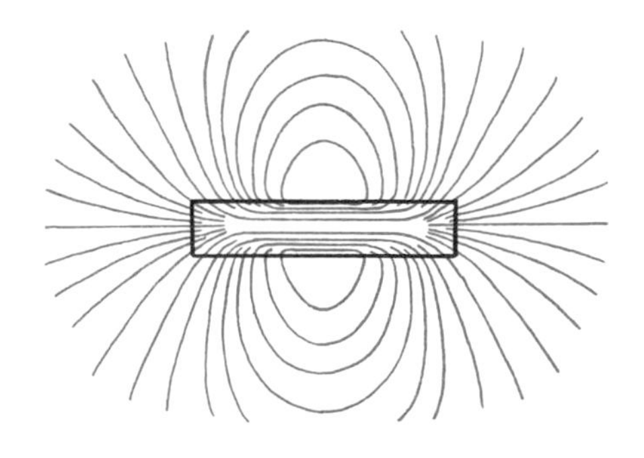

all lines complete their paths inside the magnet

Figure 2-13 Lines-of-Force Map, Bar Magnet

Figure 2-13 is a lines-of-force map based on the iron-filings pattern of a bar magnet. The lines of force form unbroken loops that stretch from pole to pole and complete their paths inside the magnet. The lines never touch or cross but actually repel each other. Only the force lines close to the magnet can be traced in full. Farther away, the magnetic force is too weak to align the iron filings; but it is still there. Modern electronic detection devices can trace the presence of a bar magnet's force over a distance of many yards.

If it were possible to show all the lines of force of a magnet at the same time, we would see its **field of force**. This field is

Figure 2-14 Field of Force, Bar Magnet

the region around the entire magnet in which its force can be detected. Our experiments with iron filings show only a single slice of that field.

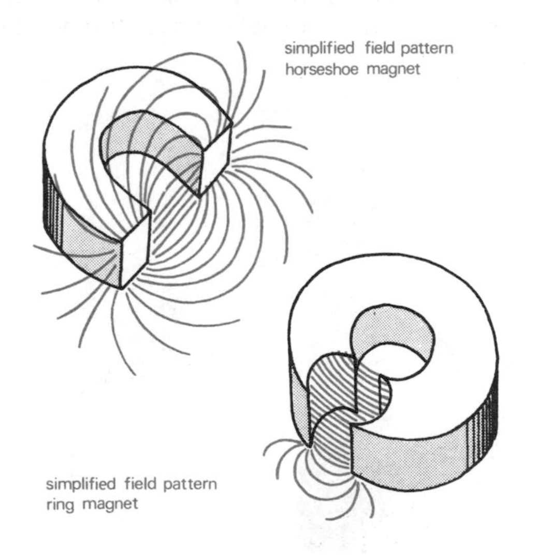

Figure 2-15 Field of Force, Horseshoe and Ring Magnet

2-6 CHARACTERISTICS OF MAGNETIC FORCE

The behaviour of magnetic force can be shown with a few simple experiments. If you wish to try these experiments, refer to chapter 17 for additional information.

2-6A The Law of Magnetic Poles

If you bring the N-pole of a bar magnet close to the N-pole of another magnet hanging from a string, both poles will

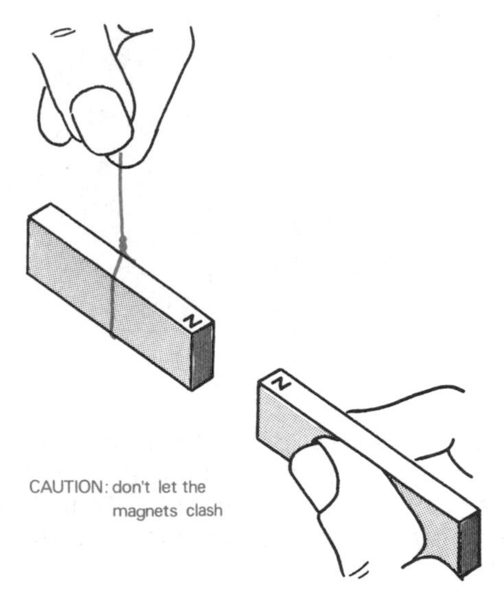

Figure 2-16 Proving the Law of Magnetic Poles

strongly repel each other. If you turn the hand-held magnet around and bring its S-pole close to the N-pole of the other magnet, the two poles will attract one another. This experiment proves the law of magnetic poles:

Like poles repel
Unlike poles attract

2-6B The Behaviour of Flux Lines

If you place two bar magnets end to end on a wooden surface, separate them with a flat eraser, cover them with a sheet of paper, and sprinkle iron filings on the paper, the resulting pattern of the iron filings will show you the behaviour of the flux lines.

When **like poles** face each other, the opposing flux lines bend away sharply. Their repelling force tends to push the magnets apart.

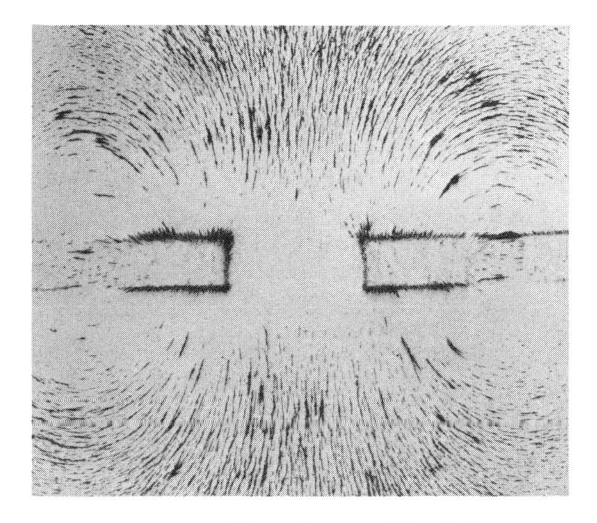
Figure 2-17 Iron-Filings Pattern, Like Poles

When **unlike poles** face each other, the flux lines stretch from one pole through the eraser to the other pole. Their combined forces tend to pull the magnets together.

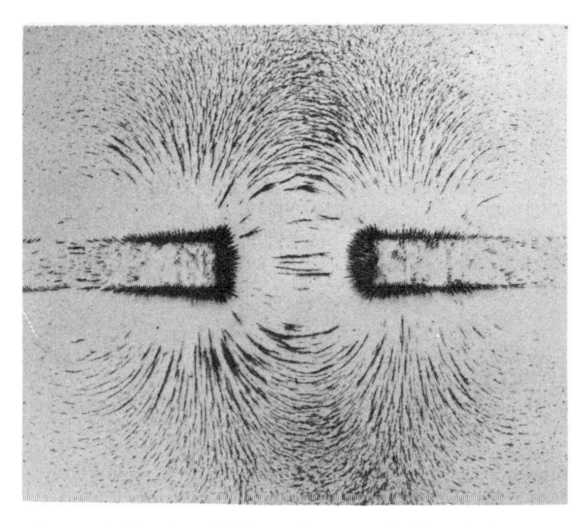
Figure 2-18 Iron-Filings Pattern, Unlike Poles

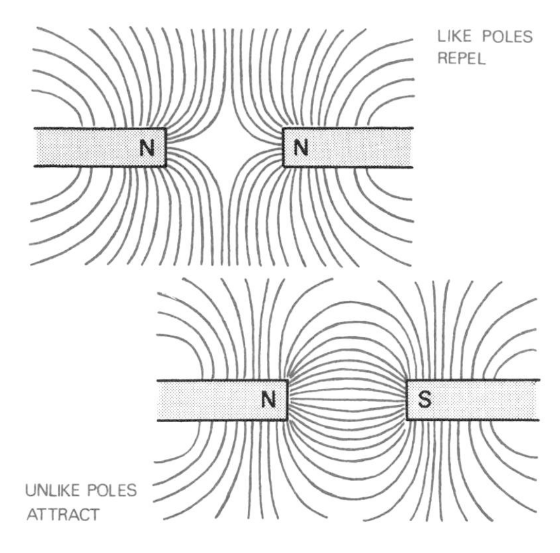

Figure 2-19 Lines-of-Force Maps — The Law of Magnetic Poles

2-6C The Direction of Magnetic Force

Magnetic force acts in a definite direction, from the N-pole toward the S-pole in the space surrounding a magnet.

If a bar magnet is held firmly on a wooden table, and if a second, loosely held magnet is brought close to it, the firmly held magnet will move the second magnet.

Figure 2-20 Demonstrating the Direction of Magnetic Force

By **international agreement**, magnetic force is said to act in the direction in which it pushes an N-pole which is free to move. The magnetic force itself is non-moving, or static. It is the free N-pole that moves.

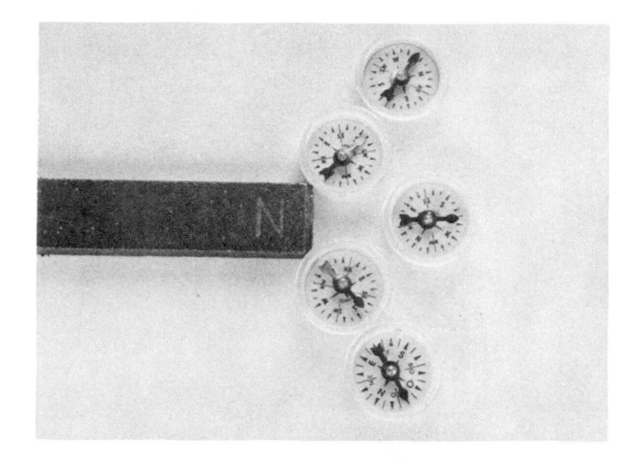

Figure 2-21 Showing the Direction of Magnetic Force

The direction of magnetic force can also be shown by placing a number of small magnetic compasses around a bar magnet. The N-poles of all the little compass needles are pushed away by the magnetic force of the bar magnet, and all point toward the magnet's S-pole.

2-6D Magnetic Force vs. Distance

The more closely a magnetic object approaches a magnet, the more strongly the magnet's force will act on it. Each time the distance between magnet and object is cut in half, the magnetic force becomes **four times stronger**.

2-7 MAGNETS IN SERIES AND IN PARALLEL

Individual magnets can be linked together in various ways to produce a stronger magnetic force. Depending on the physical arrangement of the coupled magnets, we speak of them as magnets in series or magnets in parallel.

2-7A Magnets in Series

When two or more bar magnets of **identical** size are attached to each other with unlike poles facing, their flux lines **increase in length**; but the number of lines remains that of a single magnet. If, for example, three magnets with 10,000 flux lines each are connected in series, the longer series magnet still has only 10,000

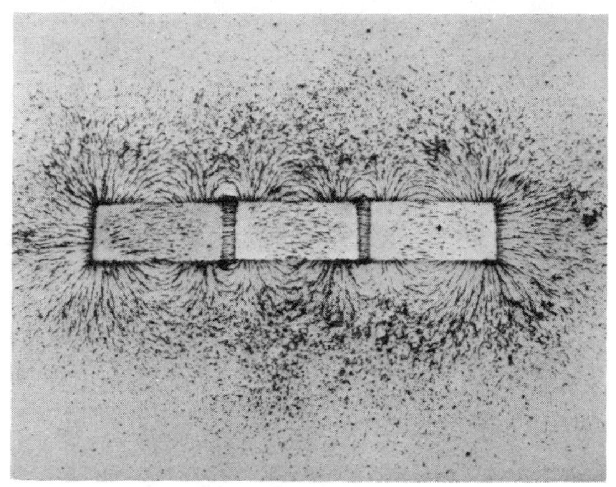

Figure 2-22 Magnets in Series—Filings Pattern

flux lines. Since a free N-pole would be pushed along a three-times-longer path than with a single magnet, the **magnetic energy** of the series magnet is three times greater.

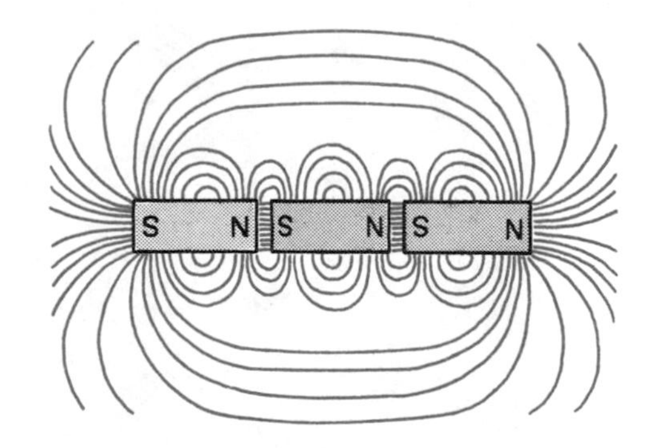

Figure 2-23 Magnets in Series—Lines-of-Force Map

Ring magnets are made of special materials pressed into the shape of tubular rings called **toroids**. Their entire magnetic force acts inside the ring. Such magnets play an important role in the memory system of computers and serve as cores for special transformers.

Figure 2-24 Ring Magnets

If small bar magnets are arranged in the shape of a ring, as in Figure 2-25, their magnetic force will be similar to that of a ring magnet. Although most flux lines will act inside the magnets, many will "leak" into the surrounding space where the magnets are joined. This leakage of flux weakens the force of this crude ring magnet. In true ring magnets, all flux lines stay inside the metallic ring.

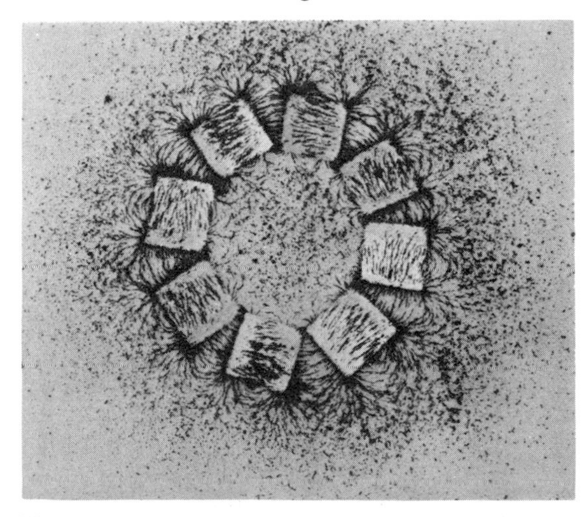

Figure 2-25 Magnets in Series—Ring

2-7B Magnets in Parallel

Magnets are arranged in parallel, or are laminated, to provide a greater number of flux lines than a single magnet. Such a laminated magnet has the combined lift-

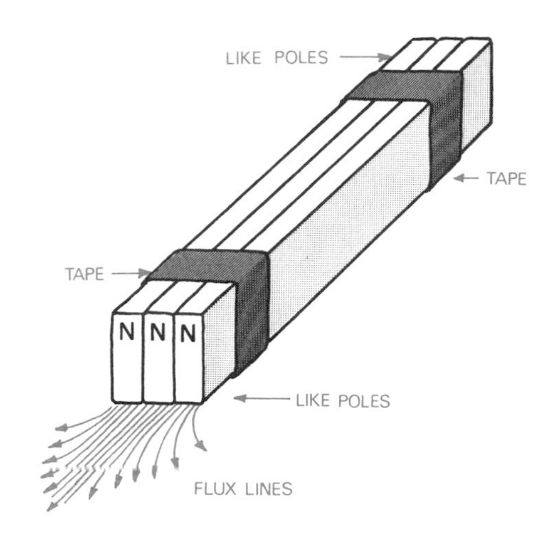

Figure 2-26 Laminated Magnet

ing power of its component magnets. If, for example, three magnets having 10,000 flux lines each are laminated together, the resulting parallel magnet has 30,000 flux lines. Its magnetic energy is three times that of a single magnet.

2-8 NON-MAGNETIC AND MAGNETIC MATERIALS

Materials which are attracted by a magnet are called **magnetic materials**. Sometimes the term ferromagnetic materials is used, because they behave like iron (Latin name: ferrum) in the field of a magnet.

Magnetic materials are **iron, nickel, cobalt**, and their various alloys. All substances that do not contain one of these three metals in some form are not attracted by a magnet. These are the **non-magnetic** materials; their list is nearly endless.

The behaviour of magnetic and non-magnetic materials in the field of a magnet can be shown with a few simple experiments.

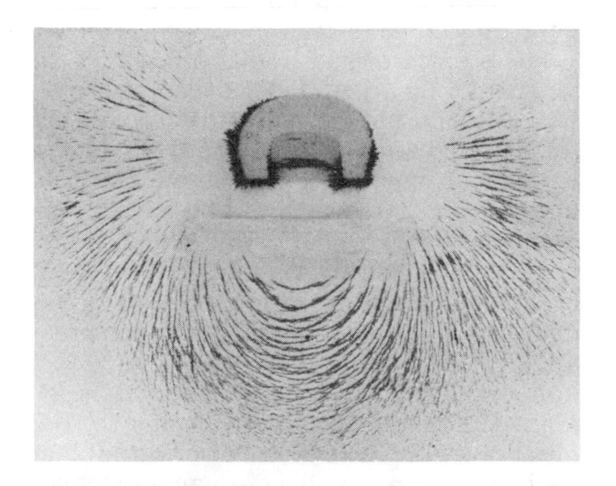

Figure 2-27 Iron Fillings Pattern of Non-Magnetic Material in Magnetic Field

If you place a piece of non-magnetic material, such as copper or aluminum, across the poles of a horseshoe magnet, cover the assembly with a sheet of paper, and make an iron-filings pattern, the result will be similar to Figure 2-27. The flux lines extend through the non-magnetic material without changing their paths. Flux lines act in non-magnetic materials exactly as they do in air.

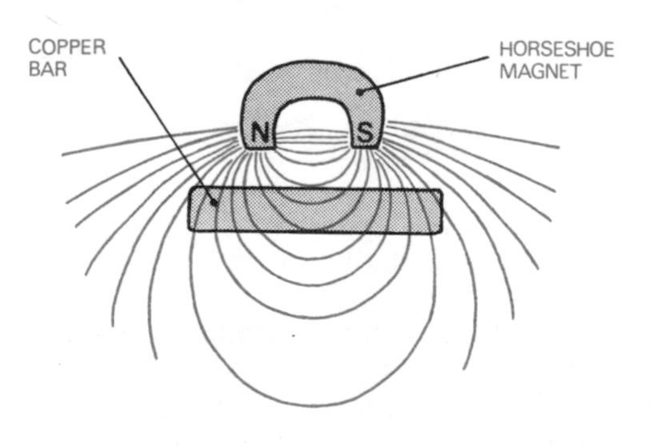

Figure 2-28 Lines-of-Force Map of Non-Magnetic Material in Magnetic Field

If you place a piece of magnetic material, such as iron, across the poles of the horseshoe magnet, separate it from the poles with a flat eraser, and make an iron-filings pattern, the result will be similar to Figure 2-29.

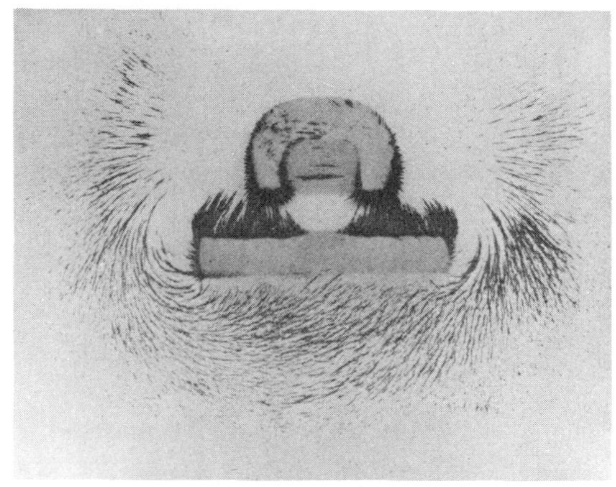

Figure 2-29 Iron-Filings Pattern of Magnetic Material in Magnetic Field

The flux lines penetrate the non-magnetic eraser as if it were air. The iron, however, **concentrates** the lines within itself and changes their paths to do so. This is proof that lines of magnetic force prefer to act in magnetic substances rather than in non-magnetic ones.

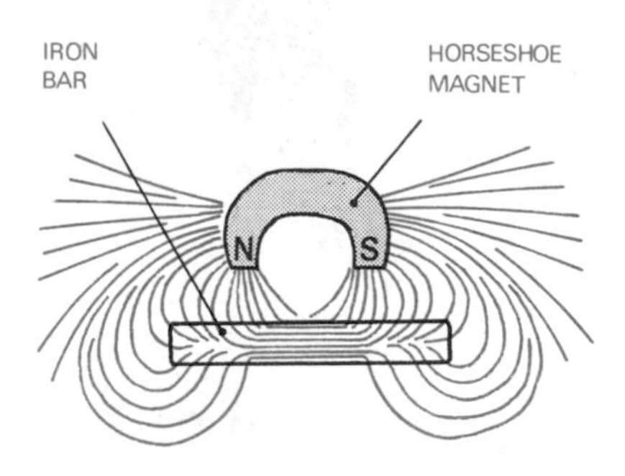

Figure 2-30 Lines-of-Force Map of Magnetic Material in Magnetic Field

For this reason, all magnetic substances are called **conductors** of flux lines, and non-magnetic materials, **non-conductors** of flux lines.

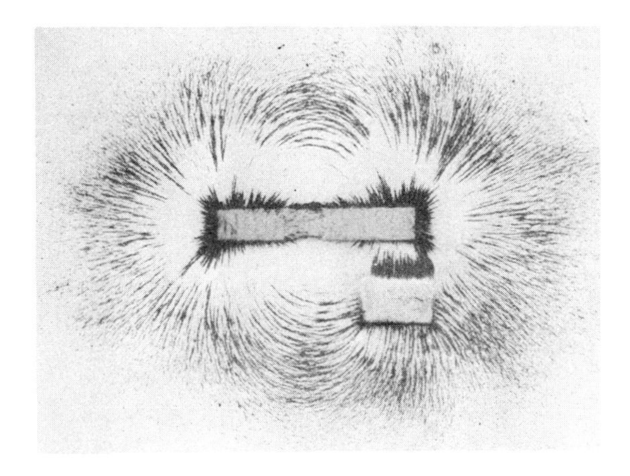

Figure 2-31 Iron as a Conductor of Flux Lines

In Figure 2-31, a small iron bar on one side of a bar magnet conducts most of the flux lines through itself, whereas the lines on the other side are unaffected. This is further proof that magnetic materials are conductors of flux lines.

2-9 INDUCED MAGNETISM

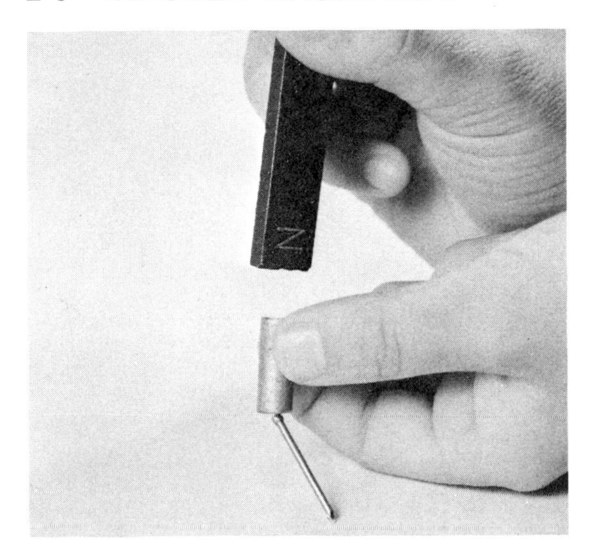

Figure 2-32 Inducing Magnetism in Soft Iron

It is possible to bring about, or **induce**, temporary magnetism in a piece of soft iron. The iron then behaves like a magnet, attracting other ferrous metals to itself.

If a piece of soft iron is brought close to a small nail, it will not attract the nail. The soft iron has no magnetic force.

If the experiment is repeated and a permanent magnet is held above the iron, the iron will attract the nail as long as the permanent magnet is nearby. The iron has become a **temporary magnet**.

What happens inside the iron that causes it to attract the nail only in the presence of a nearby permanent magnet? Soft iron consists of billions of iron atoms arranged in a regular pattern similar to that of all metals. Scientists discovered that each iron atom is a tiny permanent magnet with an N-pole and an S-pole.

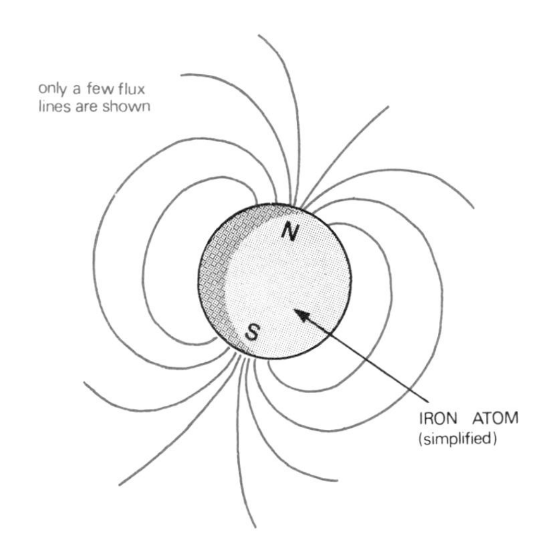

Figure 2-33 Magnetic Iron Atom

In soft iron, these **atomic magnets** can swivel rather easily. Normally, they form many closed magnetic loops that trap all the flux lines within themselves, just like a ring magnet. As a result, no magnetic force can be detected around soft iron.

When the permanent magnet comes close to the soft iron, its strong magnetic force pulls the atomic magnets out of their

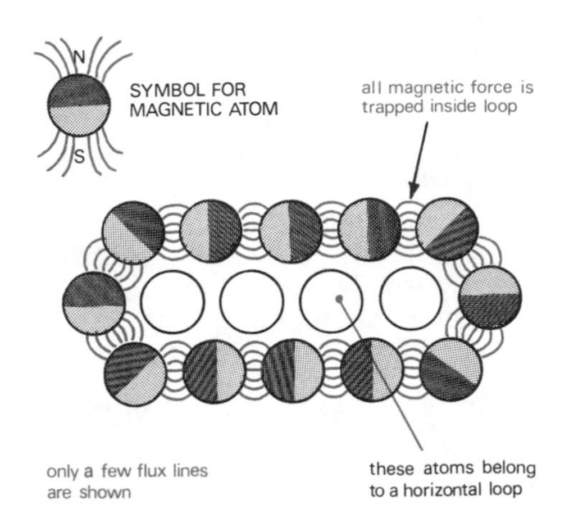

Figure 2-34 Magnetic Loops in Soft Iron

closed-loop alignment. Most of them now point in the same direction and combine their tiny magnetic forces into one strong magnetic field, as shown in Figure 2-35. This field attracts the nail.

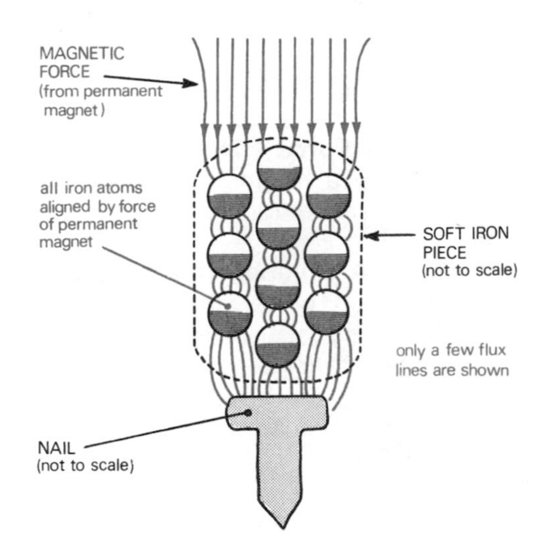

Figure 2-35 Induced Magnetism in Soft-Iron— Alignment of Atomic Magnets

When the permanent magnet is removed from the soft iron, the atomic magnets swivel back to their former closed-loop alignment, leaving the iron without external magnetic force, and the nail falls away. Since soft iron exerts a magnetic force only in the presence of a permanent magnet, its magnetism is called **temporary magnetism**.

Similar actions take place in the nail. Its atomic magnets also are pulled out of their normal closed-loop alignment, and the nail also becomes a temporary magnet.

2-10 PERMANENT MAGNETS

The atomic magnets in hard steel do not swivel as easily as those in soft iron. It is therefore more difficult to induce magnetism in steel, and we must usually bring the permanent magnet in direct contact with it. Also, once the atomic magnets are aligned in a given direction, most of them cannot return to their former closed-loop alignment without outside help. Therefore, if hard steel or other ferromagnetic alloys of great hardness are magnetized, they become **permanent magnets**.

If the N-pole of a strong bar magnet is stroked along the shank of a steel screwdriver, as shown in Figure 2-36, the shank will become a permanent magnet. The magnetic pole at the blade end of the screwdriver will be an S-pole. Can you explain why this is so?

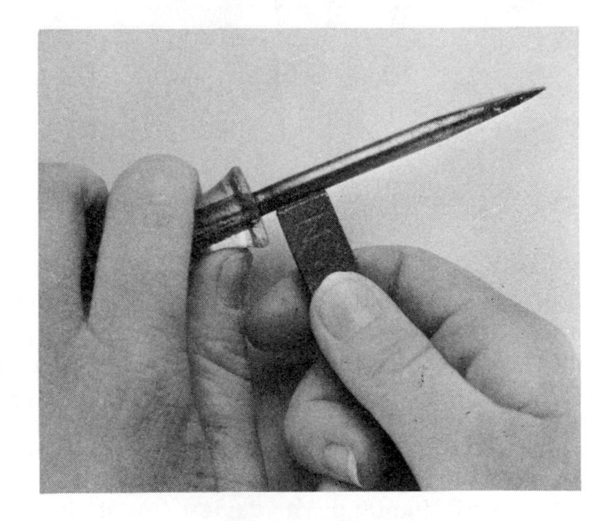

Figure 2-36 Magnetizing a Screwdriver

Most permanent magnets are made with a special alloy called **alnico**, a combination of aluminum, nickel, and cobalt. (The aluminum atoms serve as "spacers" in the metal's structure.) Alnico magnets are extremely powerful and very brittle. They break easily and lose much magnetic force if handled roughly. They are cast into many shapes and sizes and are magnetized by powerful electromagnets. Alnico magnets keep their magnetism indefinitely if properly handled.

2-11 MAGNETIC SHIELDS

Since all substances are permeated by flux lines, the shielding of delicate instruments against unwanted magnetic force poses a special problem.

Highly accurate clocks, precision meters and other instruments are adversely affected by magnetism. Such instruments are protected by enclosing them in **soft-iron magnetic shields**.

Figure 2-37 Magnetic Shield Protecting a Compass

Soft iron is an excellent conductor of flux lines. Therefore, flux lines will not penetrate the iron to reach the inside of the shield, but stay within the soft iron if the shielding is thick enough. Figure 2-37 shows that a magnetic compass inside a soft-iron shield is nearly unaffected by a powerful magnetic field. For complete magnetic shielding, a device must be **totally enclosed** in soft iron.

2-12 TECHNICAL TERMS USED IN WORKING WITH MAGNETISM

Several technical terms are essential to your further study of magnetism. They are introduced here only to make you familiar with their general meaning.

Magnetic Circuit A metallic pathway for flux lines, usually consisting of soft iron. A practical magnetic circuit for an electric instrument is shown in Figure 2-38.

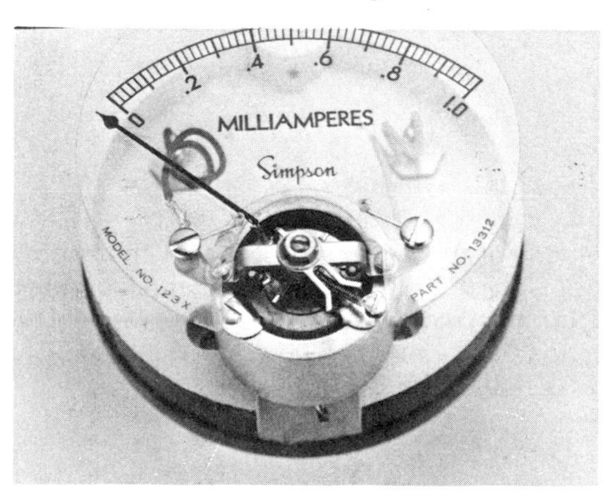

Figure 2-38 Practical Magnetic Circuit

Magnetomotive Force The force that sets up flux lines in a magnetic circuit. It can be a permanent magnet or an electromagnet.

Permeability A number rating that indicates how much better than air a given material will conduct flux lines. If a certain magnetomotive force sets up 20 flux lines in an air space, but 5,000 lines in a soft-iron piece of equal size, the permeability of the iron is 250.

Magnetic Saturation This is the condition reached when all magnetic atoms in a material are aligned in the same direction. A further increase in the magnetizing force acting on the material will not produce additional flux lines.

Retentivity The ability of a material to retain flux lines when the magnetizing force is removed. The harder the material, the greater is its retentivity.

Residual Magnetism The left-over magnetism in soft iron when the magnetizing force is removed. The greater the retentivity of a magnetic material, the more residual magnetism will exist in it.
Reluctance The opposition of a magnetic material to the setting up of flux lines. The greater the reluctance of a magnetic material, the fewer flux lines a given magnetizing force will produce in it.

2-13 ASSIGNMENTS: MAGNETS AND MAGNETISM

2-13A Write Full Answers

1. Define the terms magnet and magnetism.
2. Name the three main types of magnets.
3. From where do the terms magnet, magnetic, and magnetism come?
4. Why are natural and artificial magnets called permanent magnets?
5. State Gilbert's main discovery in his work with magnetism.
6. Describe the action of the earth's magnetic force on a lodestone compass.
7. Define accurately the terms north magnetic pole and north geographic pole, and state their relationship.
8. Draw a labelled sketch of a magnetic compass.
9. Explain carefully how a magnetic compass is oriented.
10. State the main characteristics of lines of magnetic force.
11. Draw a neat lines-of-force map for a bar magnet.
12. What is meant by the field of force of a magnet?
13. Explain the difference between lines of force and a field of force.
14. State the law of magnetic poles.
15. Make neat lines-of-force diagrams showing how flux lines behave when a) like poles face each other b) unlike poles face each other.
16. State the international rule for the direction of magnetic force.
17. Describe the relationship between strength of magnetic force and distance.
18. Why are magnets joined in series?
19. What are the effects of arranging magnets in parallel?
20. Explain the meaning of the term leakage flux.
21. Name the characteristics of a) magnetic materials b) non-magnetic materials.
22. Define the term induced magnetism.
23. Describe what happens in soft iron when magnetism is induced in it.
24. What metals are alloyed together in alnico magnets?
25. Explain the relationship between magnetomotive force and the number of flux lines in a magnetic circuit.
26. How are retentivity and residual magnetism related?

2-13B Indicate True or False

1. A lodestone is a natural magnet.
2. An electromagnet retains its magnetism indefinitely.
3. The magnetic force of a permanent magnet can be controlled.
4. The north magnetic pole is located at the geographic north pole of the earth.
5. The coloured end of a compass needle is its northseeking pole.
6. The card of a compass is divided into 360 equal degrees.
7. A magnetic compass always points exactly north.
8. The magnetic force of a bar magnet can be detected at a distance of several yards.
9. Flux lines begin at the N-pole and end at the S-pole of a magnet.
10. The field of force of a magnet acts only near its poles.
11. Flux lines coming from unlike magnetic poles join.
12. The direction of magnetic force depends on the point of view of the experimenter.
13. Magnetic force is doubled when the distance through which it acts is cut in half.

14. Magnets in series have longer flux lines than a single magnet.
15. Magnets in parallel have the combined lifting power of the individual magnets.
16. Flux leakage occurs only in completely closed ring magnets.
17. There are thousands of magnetic substances and only a few non-magnetic ones.
18. Lines of magnetic force act in non-magnetic materials exactly as in air.
19. Non-magnetic materials conduct lines of magnetic force.
20. Atomic magnets in soft iron can easily be aligned in various directions.

2-13C **Select Correct Answer**

1. A natural magnet is a piece of
   a) pure iron
   b) hard steel
   c) iron ore
   d) metal alloy
2. The scientific name for naturally magnetic ore is
   a) magnetite
   b) Stones of Magnesia
   c) lodestone
   d) iron ore
3. A bar magnet hanging from a string points toward
   a) the geographic north pole
   b) the geographic south pole
   c) the centre of the earth's magnetism
   d) the north magnetic pole
4. N-pole is an abbreviation for
   a) north pole
   b) northseeking pole
   c) north magnetic pole
   d) north geographic pole
5. South-west as a direction can also be stated as
   a) 225°
   b) 150°
   c) 180°
   d) 135°
6. The poles of a bar magnet are located
   a) at its centre
   b) exactly at the ends of the magnet
   c) a small distance from its centre
   d) at spots near each end
7. An alternate name for lines of magnetic force is
   a) magnetic lines
   b) flux lines
   c) force lines
   d) field lines
8. Flux lines
   a) stop at the poles of a magnet
   b) form closed loops
   c) exist only in one plane
   d) cross over inside the magnet
9. The field of force of a magnet
   a) can be made visible in its entirety
   b) acts only between the poles of the magnet
   c) is weakest at the poles of the magnet
   d) exists in the entire space surrounding the magnet
10. Flux lines coming from like magnetic poles
    a) repel and bend away sharply
    b) join
    c) pull the poles together
    d) do not act on one another
11. Magnetic force acts in the direction
    a) from S-pole to N-pole outside the magnet
    b) from N-pole to S-pole inside the magnet
    c) in which it pushes a free N-pole
    d) in which it pulls a piece of iron
12. The strength of magnetic force
    a) remains the same at all distances
    b) decreases as an object comes closer to it
    c) increases fourfold each time the distance is halved
    d) increases as the object moves farther away
13. A magnet attracts an iron object with a force of 10 pounds at a distance of 1 inch. At a ½-inch distance, the strength of the force will be
    a) 40 pounds
    b) 20 pounds
    c) 10 pounds
    d) 5 pounds

14. Magnets in series
    a) oppose one another
    b) tend to cancel their magnetic fields
    c) have more flux lines than a single magnet
    d) have longer flux lines than a single magnet
15. Magnetic materials
    a) contain no ferromagnetic substances
    b) are conductors of flux lines
    c) include all metals
    d) are not penetrated by magnetic force
16. The atomic magnets in hard steel
    a) swivel easily
    b) require much force for realignment
    c) do not retain their alignment
    d) always return into closed loops
17. The atoms of all magnetic substances
    a) are not affected by magnetic force
    b) repel flux lines
    c) have no magnetic force of their own
    d) are tiny permanent magnets
18. The permeability of a magnetic substance is the material's
    a) ability to conduct flux lines
    b) opposition to magnetic force
    c) ability to retain flux lines
    d) total magnetic field
19. The reluctance of a magnetic material is its
    a) total number of lines of force
    b) ability to hold flux lines
    c) opposition to the setting up of flux lines
    d) ability to retain flux lines
20. Magnetic saturation
    a) occurs in non-magnetic materials
    b) depends on a material's retentivity
    c) removes flux lines from a substance
    d) is reached when a substance contains all the flux lines it can hold

3

BASIC FACTS ABOUT ELECTRICITY

Electricity was unknown until the early seventeenth century. At that time William Gilbert, whom we encountered in our study of magnetism, began to investigate ancient Greek accounts of a mysterious force of attraction in amber that had been rubbed with fur.

Figure 3-1 Amber Attracting Chaff

Gilbert discovered that many other materials when rubbed with fur or silk attract light objects in the same way as amber. Although he did not pursue the subject any further, he did coin a new word. He said that the materials had become **electrified**. Gilbert based this new term on the Greek word for amber, **elektron**.

3-1 STATIC AND DYNAMIC ELECTRICITY

If you walk across a nylon carpet on a dry day, the friction between the soles of your shoes and the carpet fibres will build up an **electric charge** on your body. This charge is called **static electricity**, because it remains static, or stationary, on your body until you touch some metallic object.

At the instant you touch a doorknob or other metal surface, the electric charge **flows** from your body into the metal. This flowing or moving charge is called **dynamic electricity**. It can give you a slight electric shock, and the discharge is usually followed by a snapping sound. Sometimes you can see a bluish spark.

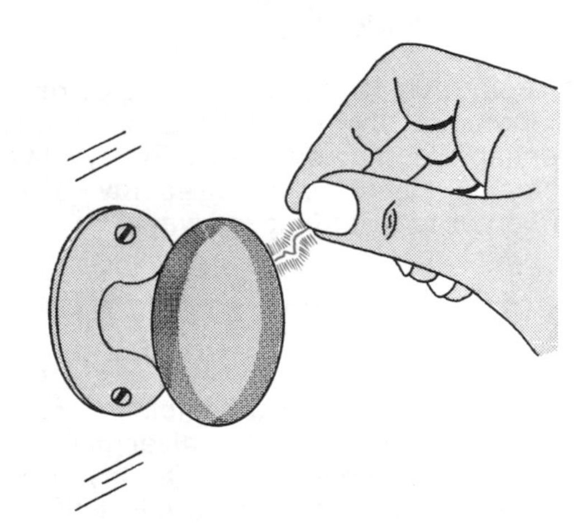

Figure 3-2 Discharge Spark

Static electricity is nearly always generated by friction between two objects, or by separating materials that have been in close contact. It can build up only on objects that are **insulated** from their surroundings. If a path exists, the electricity generated by the friction will drain off into the earth without producing a static charge.

Dynamic electricity needs a **conductor** to flow through. If large static charges have accumulated on an object, the surrounding air becomes a conductor, and the charge rushing through it will produce an intense bluish arc. Lightning is the most awesome example.

Figure 3-3 Lightning

Both static and dynamic electricity are **forms of energy**. Static electricity is potential electric energy, and dynamic electricity, kinetic electric energy. They can be put to work in many ways.

3-2 THE INTERNAL STRUCTURE OF SOLID MATERIALS

The way in which atoms combine to form solids determines whether a given material is an electrical conductor or an insulator. The internal structure of materials is therefore of tremendous importance to electrical technology.

3-2A Atoms and Molecules

An atom is the smallest part of a pure substance (element) that still has all the characteristics of that substance. Some 105 elements are known at this time. Therefore, 105 different kinds of atoms make up all the materials around us, including ourselves.

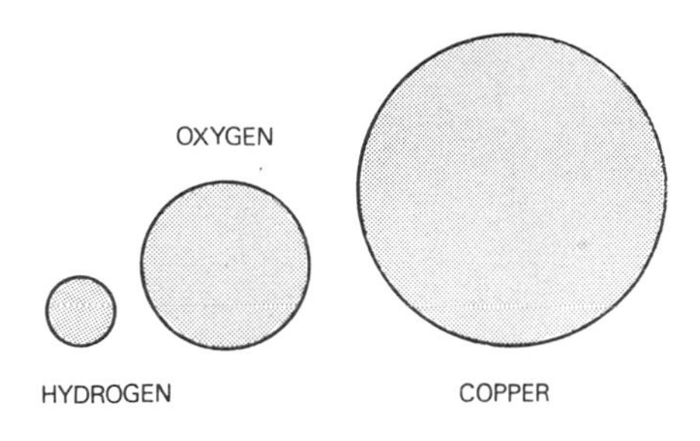

Figure 3-4 Relative Size of Atoms

All atoms have a similar structure, but differ in size and in weight. All atoms except ordinary hydrogen are composed of three elementary particles: **protons, neutrons** and **electrons**. The protons and neutrons are always found in the centre of the atom, where they form a small, dense and heavy **nucleus**. The electrons, the smallest and lightest of the elementary particles continuously circle the nucleus, forming an **electron cloud.**

BASIC ATOMIC STRUCTURE
(Model – not to scale)

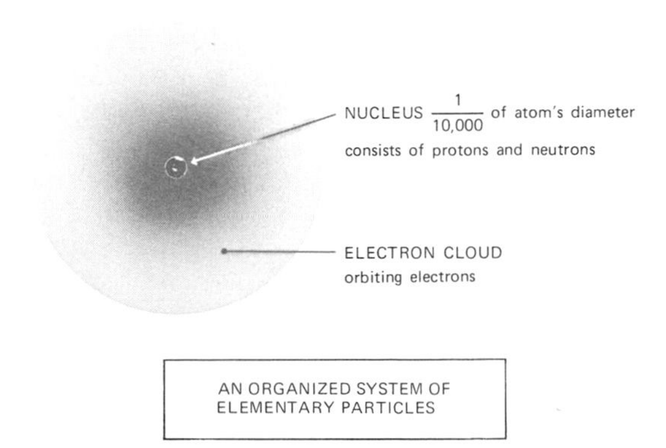

AN ORGANIZED SYSTEM OF ELEMENTARY PARTICLES

Figure 3-5 The Basic Structure of All Atoms

Many pure substances are composed of seemingly endless rows of identical atoms stacked row upon row, with small areas of empty space between them. Such an arrangement is called a **crystal structure**. It is typical of the solid state of many materials.

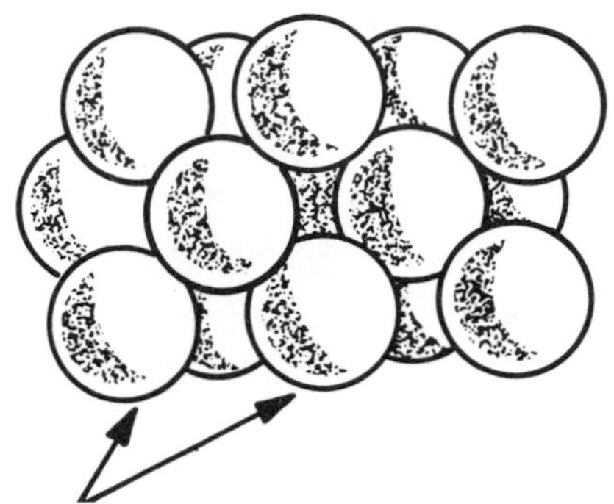

Figure 3-6 Atoms in a Crystal Structure

Mixed substances or **compounds** consist of two or more different kinds of atoms. In such materials, the different atoms first combine into clusters called **molecules**. The molecules, in turn, form the solid.

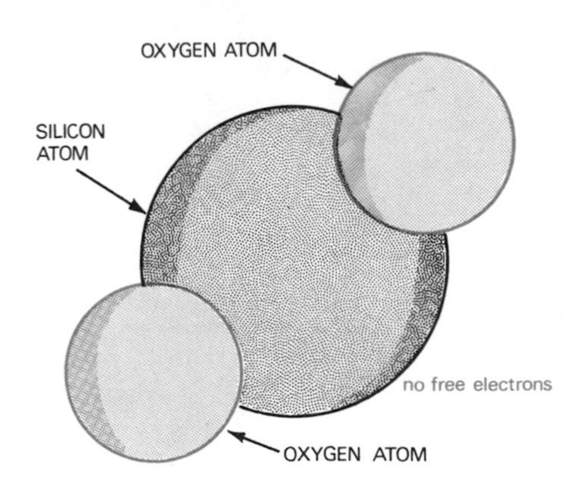

Figure 3-7 Glass Molecule

3-2B The Structure of Metals

More than half of the 105 elements are metals. Some of these are very important to electrical technology, especially silver, copper, aluminum, gold and tungsten.

All pure metals consist of billions and billions of identical atoms packed as tightly as possible. But the space between the atoms is not empty. Extremely small particles called **free electrons** move through it in an irregular fashion. It is the presence of these free electrons which makes all metals **conductors** of electricity, although not all conduct equally well.

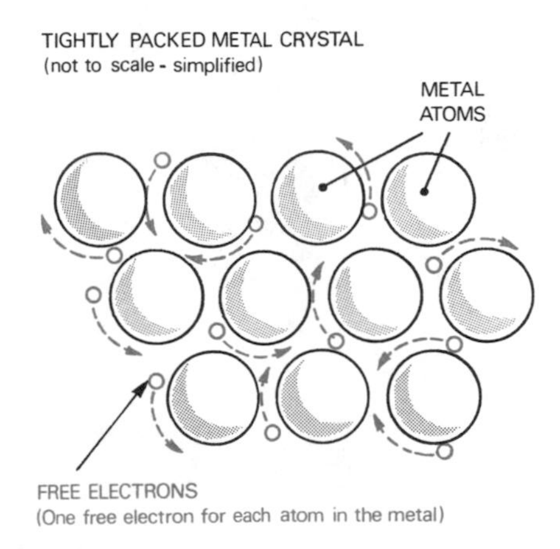

Figure 3-8 Free Electrons in a Metal

3-2C The Structure of Non-Metals

Many non-metallic materials have an internal structure quite different from that of the metals. Their atoms team up into molecules and these, in turn, form the material. The most important point about the non-metals is the **absence of free electrons** in the relatively large spaces between the molecules. This feature makes many non-metals excellent electrical **insulators.** Some of the more important non-metallic materials are glass, porcelain, ceramics, rubber and plastics.

3-3 ELECTRIC CHARGES IN THE ATOM

Scientists have discovered that two of the three elementary particles that make up an atom possess a **permanent electric charge.** Although the nature of these charges remains a mystery, we know a great deal about them, how they work and what they can do.

Each **proton** in the nucleus has a permanent **positive charge**. This charge sends out invisible lines of electric force that **attract electrons.**

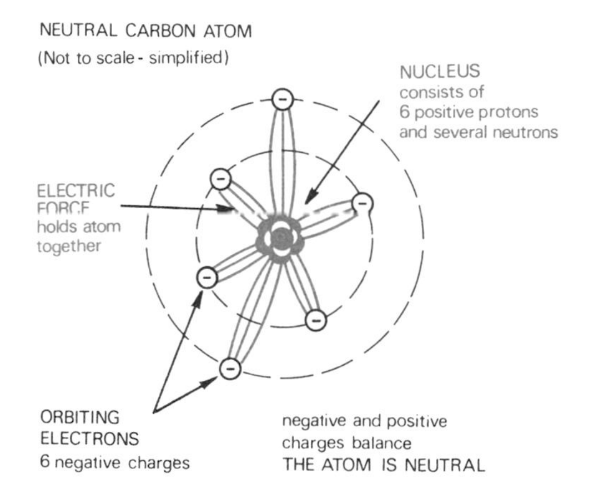

Figure 3-9 A Neutral Carbon Atom

Each **electron** in the electron cloud has a permanent **negative charge** which is of the same size as the charge on the proton, but of opposite character. It sends out invisible lines of electric force that **attract protons.**

This mutual attraction between protons and electrons holds the atom together. Each atom has exactly the same number of protons and electrons. As long as this balance is not disturbed, the atom is electrically **neutral**. It neither attracts nor repels other atoms.

The protons in the atomic nucleus cannot be removed. We can change only the number of electrons in the outer layer of the electron cloud.

If we remove one or more electrons from an atom, we upset its electrical balance. Because the protons outnumber the electrons, their invisible electric force now reaches outside the atom, trying to **attract electrons**. The atom is no longer electrically neutral; it has become a **positive ion.**

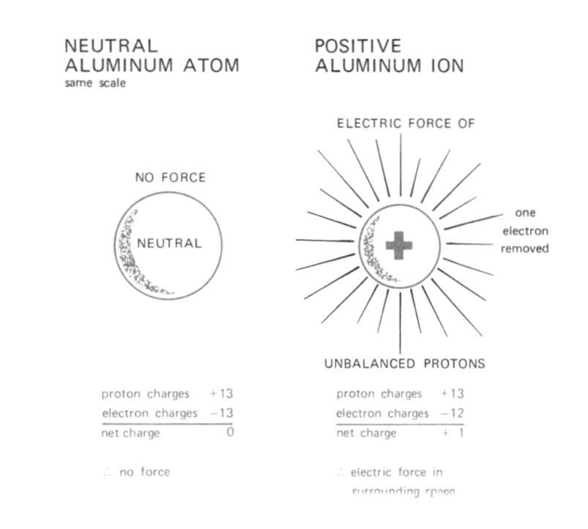

Figure 3-10 Atom and Positive Ion of Aluminum

If we force one or more extra electrons into the electron cloud of an atom, we again upset its electrical balance. The lines of electric force coming from the extra electrons reach outside the atom,

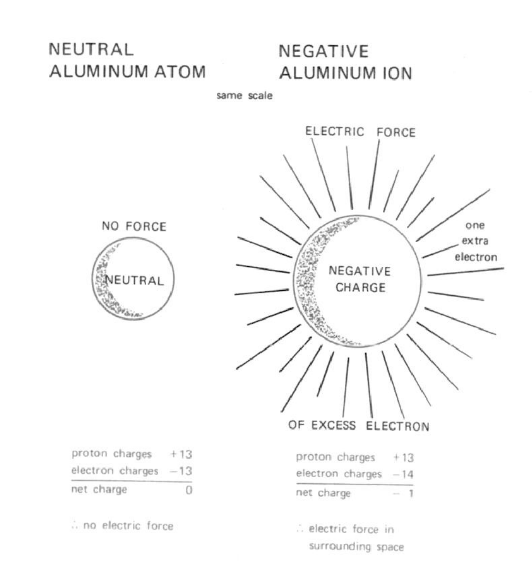

Figure 3-11 Atom and Negative Ion of Aluminum

trying to attract positive charges, or repelling other electrons. The atom is no longer neutral; it has become a **negative ion.**

Positive and negative ions attract each other exactly as protons and electrons do, inspite of their relatively large size and mass.

3-4 THE BEHAVIOUR OF CHARGED BODIES

If you run a comb through dry hair, you will give the comb a **negative** electric charge. The friction created by the comb moving through the hair **transfers** electrons from the hair to the comb. The hair itself will receive a positive charge.

If you bring the charged comb close to some small pieces of dry paper, they will cling to the comb. Some pieces will first be attracted by the comb and then

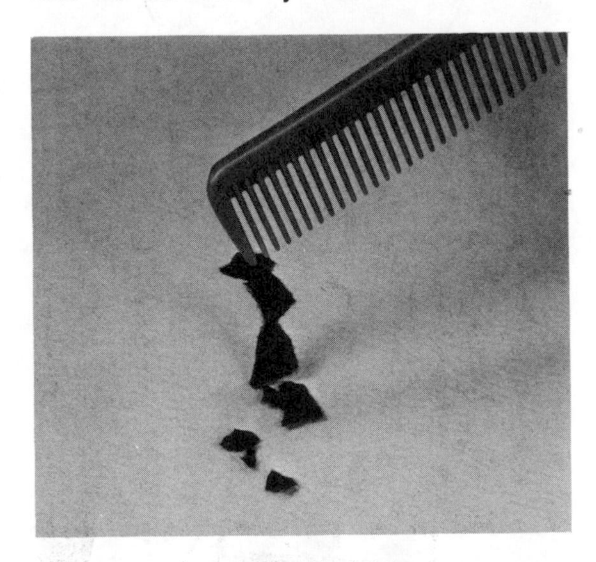

Figure 3-12 Paper Clinging to a Charged Comb

repelled. Here is what happens: the electric force coming from the negative charge on the comb attracts the neutral paper. Excess electrons move from the comb to some of the paper pieces giving them a negative charge, too. These pieces are then repelled.

This simple experiment proves the **law of charges**, which states that **like charges repel** and **unlike charges attract**.

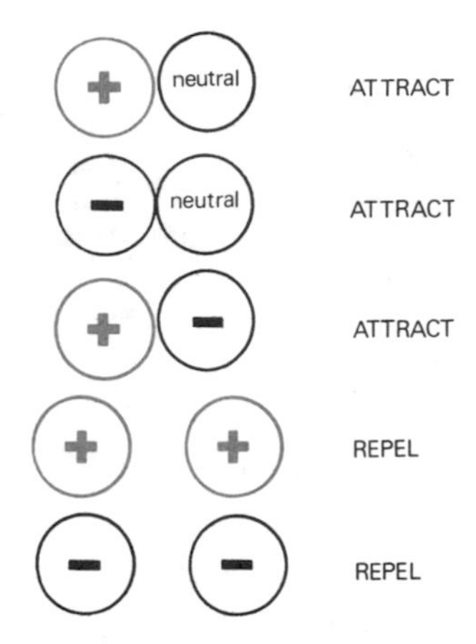

Figure 3-13 The Law of Charges

Induced electric charges are produced in all paper pieces when the charged comb approaches them. The electric force coming from the charge on the comb **displaces electrons** inside the paper, giving the side facing the comb an induced positive charge. The paper is therefore attracted by the comb.

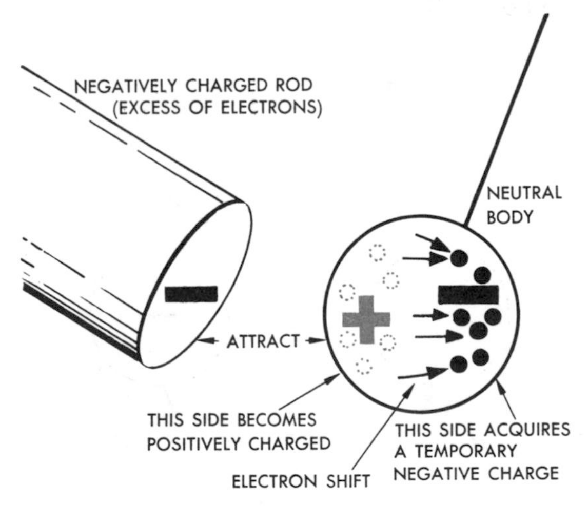

Figure 3-14A Induced Charges

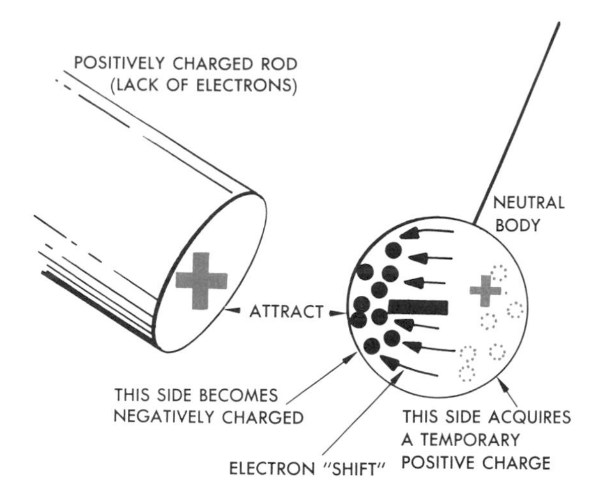

Figure 3-14B Induced Charges

A body with induced charges is still electrically **neutral** as a whole because it has neither lost nor gained electrons.

The **International (S.I.) unit of charge** is the **coulomb**. A body is said to have a charge of one coulomb if it has either lost or gained 6.25 billion billion electrons.

Since charge cannot be measured directly, there is no such thing as a coulombmeter. The charge on a body is usually determined by indirect measurements and mathematical calculations.

The symbol for charge is **Q**, and for coulomb, **C**.

3-5 VOLTAGE

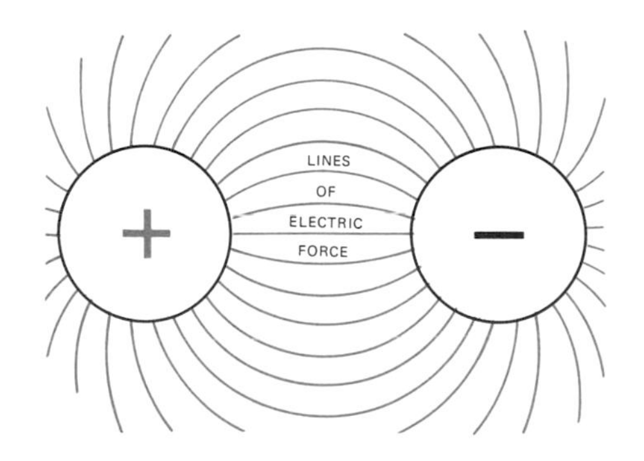

Figure 3-15 Electromotive Force Between Charged Spheres

If, by some means, electrons are forced out of one metal sphere into another, both spheres will receive opposite electric charges. These static charges create an **electromotive force** or **voltage** which **tries** to propel the excess electrons from the negative sphere back to the positive one. The greater the charges on the spheres, the stronger will be this voltage or pressure acting on the electrons.

In practical electric circuits, the voltage required to propel the free electrons through the circuit is supplied by dry cells or other power supplies.

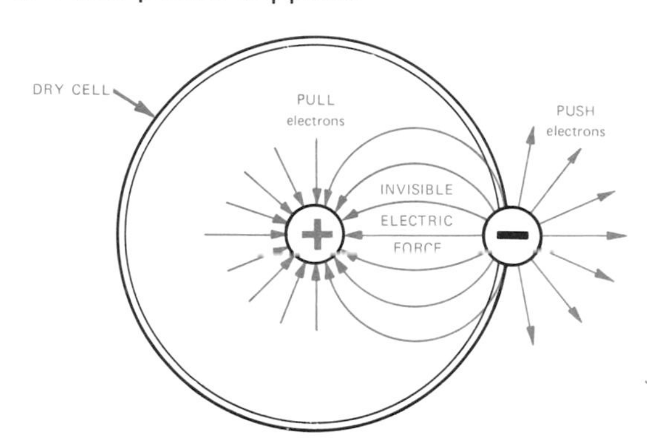

Figure 3-16 The Voltage of a Dry Cell

The **positive** terminal of the cell exerts a **pulling force** on free electrons, and the **negative** terminal, a **pushing force**. The pressure difference between the two poles is the voltage of the cell.

Figure 3-17 Using a Voltmeter

The **international (S.I.) unit of voltage** is the **volt**. A single dry cell has a voltage of 1.55 volts when new, regardless of its physical size. The voltage between the terminals of a home outlet varies from 110 to 120 volts.

The symbol for voltage is **E**, and for volt, **V**.

In practical work, voltage is measured with a voltmeter. Voltmeters come in two basic versions: continuous-scale and digital meters.

3-6 ELECTRON CURRENT

When a voltage acts on the free electrons in a metal, they are forced to **drift** or **flow** through the spaces between the metal atoms. This **electron current** always moves through the metal from the negative toward the positive terminal of the power supply.

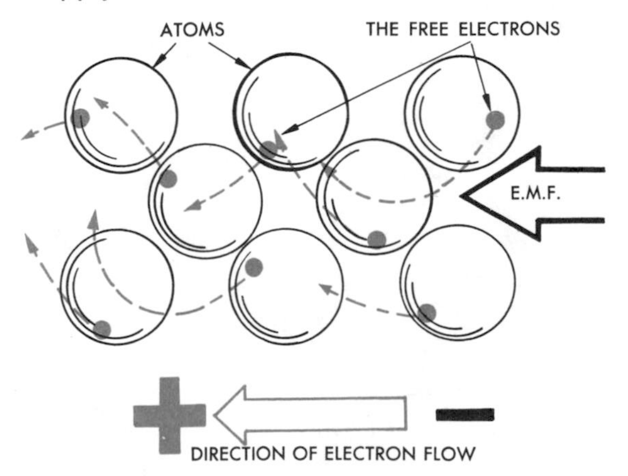

Figure 3-18 Electron Current Flowing Through a Metal

For each electron leaving the metal at the positive terminal of the voltage source, another one enters the metal at the negative terminal. Thus, so long as the voltage remains constant and the metallic path unbroken, a continuous electron current will flow.

The **international (S.I.) unit of electron current** is the **ampere**. It is defined as the flow of 6.25 billion billion electrons **per second** through the point of measurement. One ampere is equal to a charge flow of one coulomb per second.

The symbol for electron current is **I**, and for ampere, **A**.

In practical work, electron current is measured with an amperemeter, or ammeter for short. Ammeters are available in many designs.

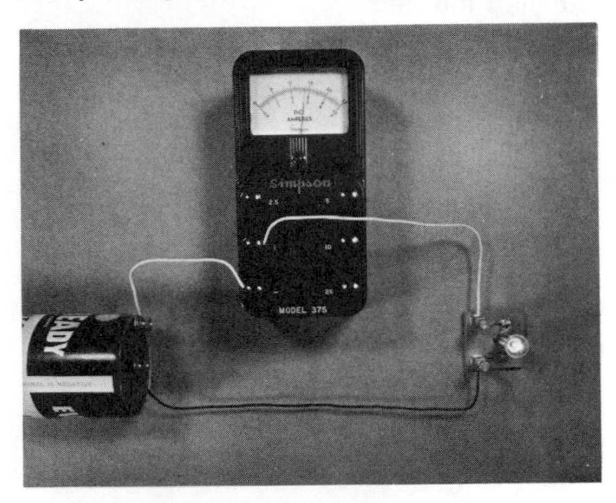

Figure 3-19 Using an Ammeter

3-7 A SIMPLE ELECTRIC CIRCUIT

An electric circuit is an arrangement of electrical components that enables us to **put electricity to work**. Each part of the circuit serves a specific purpose, and special terms have evolved for them.

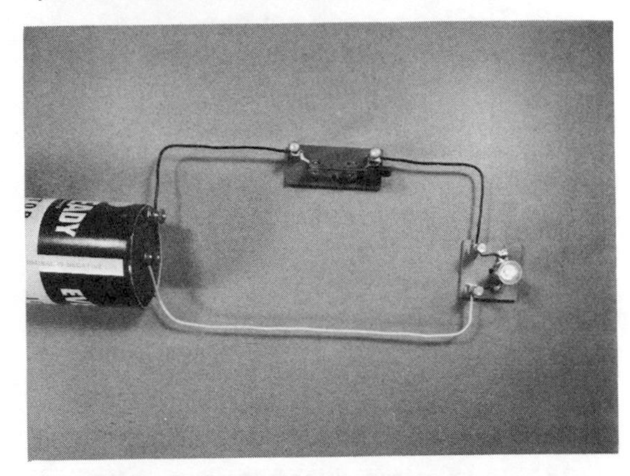

Figure 3-20 Simple Electric Circuit

The source of voltage, which gives energy to the free electrons, is called the **supply**.

The switch, which can start or stop the electron current, is called the **control**.

The insulated copper wires, which provide a low-resistance path for the free electrons, are called **conductors**.

The lamp, which is heated to a bright glow by the energy of the free electrons, is called the **load**.

SIMPLE ELECTRIC CIRCUIT
Schematic Diagram

CONTROL
(switch)

SUPPLY
(dry cell)

LOAD
(lamp)

CONDUCTORS
(insulated copper wire)

Figure 3-21 Simple Electric Circuit — Schematic Diagram

Inside the open circuit: when the knife switch is in the up position, the circuit is broken, or open, and no electron current can flow. The voltage of the supply builds up opposite charges across the terminals of the open switch. As a result, the full supply voltage will exist across the terminals. The lamp does not glow.

Inside the closed circuit: when the knife switch is pushed down, the circuit is complete, or closed. The voltage of the supply now forces an electron current to flow continuously through the entire circuit. In the copper conductors, this current meets little opposition. But in the thin filament of the lamp, the free electrons run into much **resistance**, and their kinetic energy is converted to **heat** and **light** as the supply voltage forces them through.

The basic purpose of any electric circuit is to **send electric energy** from a supply to a load, often over hundreds of miles, and to **control** this energy flow. The free electrons are the carriers of this energy.

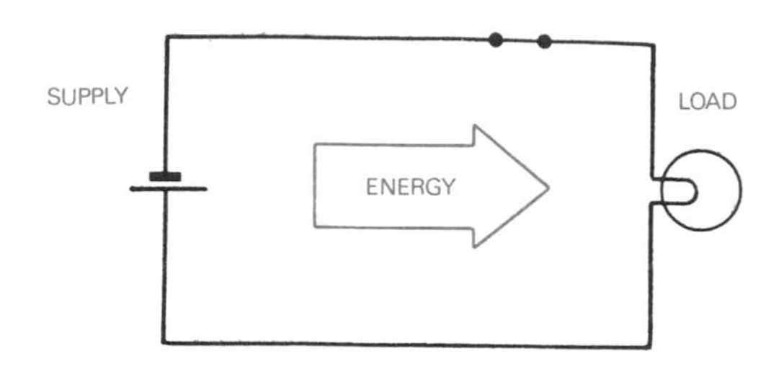

3-8 RESISTANCE

Free electrons flowing through an electric circuit encounter opposition in all parts of the circuit. This opposition to their flow is called **resistance**. It occurs when the drifting electrons **collide** again and again with the stationary metal atoms. These frequent collisions shake the atoms into vibrations that are felt as **heat** by our senses.

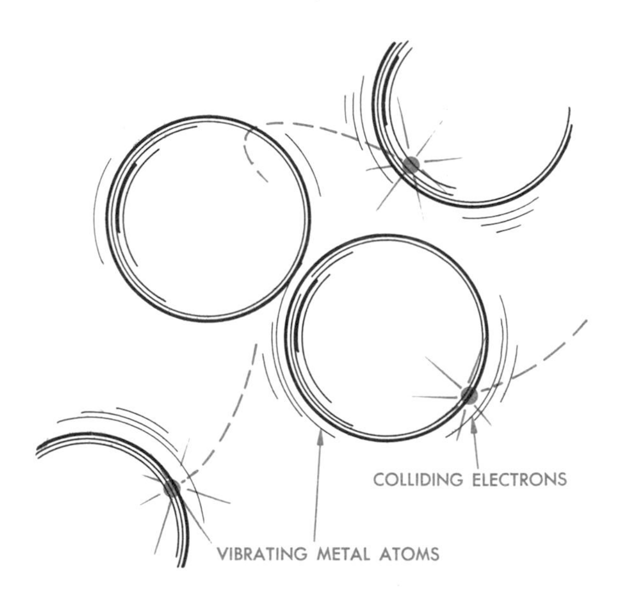

Figure 3-23 Resistance in Metallic Conductors

The load always has the greatest resistance in the entire circuit; it therefore causes the greatest **loss of electron energy**.

The **international (S.I.) unit of resistance** is the **ohm**. It is defined as that amount of resistance which requires a voltage of one volt to force a current of one ampere through it.

The symbol for resistance is **R**, and for ohm, the Greek letter omega Ω. On schematic diagrams, a resistance is represented by one of the symbols shown in Figure 3-24.

RESISTANCE SYMBOL
for electrical and electronic circuits

RESISTANCE SYMBOL
for industrial circuits

Often, a number identifies resistors in a circuit

R 204

Figure 3-24 Schematic Symbols for Resistance

In practical work, the resistance of circuit components is measured with an **ohmmeter**. This instrument can only be used **after the supply voltage has been removed** from the circuit.

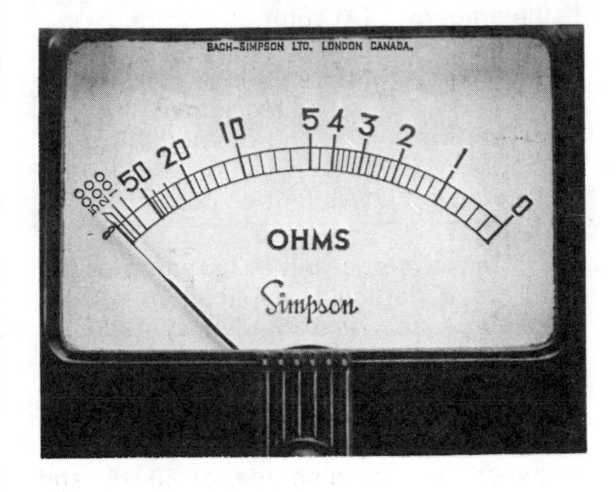

Figure 3-25 Ohmmeter

The actual resistance of a conductor depends on the following factors: the type of **material**, the **length** of the conductor, the size of its **cross-sectional area**, and its **temperature**.

The best conductor **materials** are silver, copper, gold and aluminum, in that order.

The greater the **length** of a conductor, the higher will be its resistance, because the drifting electrons encounter much more friction than in a shorter wire.

The larger the **cross-sectional area** of a conductor, the lower will be its resistance, since the metal will appear as a much wider and easier path for the electrons to flow through.

As the **temperature** of a conductor increases, so will its resistance. The higher the temperature, the more violently the metal atoms will vibrate, and the more often the flowing electrons will collide with them.

THE RELATIVE RESISTANCE OF IMPORTANT METALS

METAL	REL. RES.
Silver	0.95
Copper	1.00
Gold	1.46
Aluminum	1.64
Tungsten	3.25
Zinc	3.40
Cadmium	4.40
Nickel	5.50
Iron	5.60
Lead	12.80
Chromax	58.00

Table 3-1 Relative Resistance of Important Metals

The **relative resistance** of a material is its resistance compared to a copper conductor of equal size and at the same temperature. Table 3-1 shows some typical values.

3-9 ELECTRON CURRENT VS. CONVENTIONAL CURRENT

Today there is no longer any doubt that the electron current is the **actual method** of transporting electric energy through metallic conductors. There exists, however, an older theory of current flow based on Benjamin Franklin's ideas and still widely used.

Franklin did much experimental work with electricity. He assumed that electric charges were the result of transferring an invisible and weightless fluid from one body to another. The body losing the fluid would have a negative charge, and the body gaining it, a positive charge. It was therefore logical for Franklin to further assume that an electric current would flow from the positive back to the negative terminal of a power supply.

When Joseph Thomson discovered the existence of the electrons in 1897, he found that the charge labelled **negative** by Franklin was actually caused by an **excess of electrons**. This proved Franklin's theory of current flow wrong. However, since hundreds of books had been written by that time—all using Franklin's theory—it was decided to retain it.

Today, the term **conventional current** is used for Franklin's assumed direction of charge flow, and the term **electron current** for the actual direction of electron flow in **metallic** conductors.

3-10 ELECTRICITY AND THE METRIC SYSTEM (S.I.)

In 1960, the 11th General (International) Conference on Weights and Measures, meeting in Paris, France, adopted a new, international system of units of measurement. This new system, called the **International System of Units**, or the **S.I.**, is an expansion and refinement of the metric system which had been in international use until that time.

At the time of writing, more than ninety percent of all countries have already adopted the S.I. system, or are planning to switch over to it. The S.I. has thus become the principal system of units of measurement in the world.

The S.I. is based on **six basic units** from which all others are derived:

the **metre**	(m)	basic unit of **length**
the **kilogram**	(kg)	basic unit of **mass**
the **second**	(s)	basic unit of **time**
the **ampere**	(A)	basic unit of **electric current**
the **candela**	(cd)	basic unit of **light intensity**
the **kelvin**	(K)	basic unit of **temperature**

A unique feature of the S.I. is the naming of its units. With few exceptions, all its principal units are named in honour of researchers who made important contributions to the advancement of science and technology. Practically all electrical units belong to this group.

The **principal units of electricity** have always been metric units. They have been included in the S.I. system:

the **coulomb**	(C)	unit of electric **charge**
the **volt**	(V)	unit of electric pressure (**voltage**)
the **ampere**	(A)	unit of electric **current**
the **ohm**	(Ω)	unit of electric **resistance**
the **joule**	(J)	unit of **energy** in any form
the **watt**	(W)	unit of **power**

If a basic metric unit is too large or too small for a particular application, its size can be reduced or expanded by adding a **metric prefix** to the unit's name. Thus, 1,500 volts can also be stated as 1.5 kilovolt, or 1.5 kV, by adding the prefix kilo to the basic unit. Likewise, 0.00005 amperes becomes 50 microamperes, or 50 μA, and 0.02 volts can be expressed as 20 millivolts, and so on. The most common metric prefixes are:

Mega	(M)	multiply basic unit by 1,000,000
kilo	(k)	multiply basic unit by 1,000
milli	(m)	multiply basic unit by 1/1,000
micro	(μ)	multiply basic unit by 1/1,000,000

Many instruments are calibrated directly in expanded or reduced units. Examples are the kilovoltmeter, the megohmmeter, the millivoltmeter and the microammeter.

3-11 ASSIGNMENTS: BASIC FACTS ABOUT ELECTRICITY

3-11A Write Full Answers

1. Define the term static electricity.
2. Why can a static charge be built up only on an insulated body?
3. Explain why the internal structure of materials is so important to electrical technology.
4. With the aid of a labelled drawing, describe the basic structure of all atoms.
5. What are the chief differences in the structure of metals and most non-metals?
6. Explain why electrons and protons attract each other. Make simple drawings to illustrate your answer.
7. Describe the electrical characteristics of atoms, positive ions, and negative ions.
8. Why is a neutral piece of paper attracted by a charged comb?
9. State the law of charges.
10. Make a simple sketch showing all possible combinations of charged and neutral bodies attracting and/or repelling each other.
11. What is the relationship between charge and voltage?
12. Describe the effect of voltage on free electrons.
13. Define the international unit of electron current.
14. State the general direction of flow of an electron current.
15. Draw the schematic diagram of a simple circuit and label all its parts.
16. Explain the purpose of the four main parts of the basic electric circuit.
17. Describe what happens in the filament of a lamp when an electron current flows through it.
18. Name the four factors that affect a conductor's resistance.
19. Why did Thomson's discovery of the electrons prove Franklin's theory to be wrong?
20. Define the terms electron current and conventional current.

3-11B Indicate True or False

1. Static electricity can be built up on any object in your surroundings.
2. Both static and dynamic electricity are a form of energy.
3. The spaces between the atoms in a metal contain free electrons.
4. The free electrons in non-metals cannot be moved around.
5. Protons and neutrons form the small, dense and heavy nucleus of all atoms.
6. Protons can easily be removed from the nucleus of an atom.
7. A negative electron and a negative ion have identical size and mass.
8. The international unit of charge is the coulomb.
9. A body with induced charges has lost or gained electrons.
10. Electric charges create electromotive force.
11. Electron current always flows from the positive to the negative terminal of a voltage source.
12. Electron current is the flow of free electrons through a conductor.
13. In the electric circuit, free electrons lose most of their energy in the conductors.

14. An electron current cannot flow in a closed circuit.
15. A resistance converts the kinetic energy of free electrons to heat energy.
16. The resistance of practical circuits cannot be measured directly.
17. When using an ohmmeter, the source of voltage must first be removed from the circuit.
18. The international unit of resistance is the ohm.
19. Conventional current is the same thing as electron current.
20. Thomson's discovery of the electrons revealed the true direction of electron flow.

3-11C **Select Correct Answer**

1. Static electricity is
   a) kinetic electric energy
   b) unable to do work
   c) potential electric energy
   d) easily built up on all objects
2. Dynamic electricity
   a) can exist anywhere
   b) is the charge on a body
   c) is potential electric energy
   d) needs a conductor to flow through
3. Free electrons are found in
   a) the spaces between the atoms of a metal
   b) the electron cloud of an atom
   c) in the nucleus of the atoms
   d) in non-metallic compounds
4. Non-metals such as glass, ceramics and plastics
   a) are excellent conductors of electricity
   b) are composed of one kind of atoms
   c) are excellent electrical insulators
   d) have a tightly packed internal structure
5. A neutral atom
   a) is electrically unbalanced
   b) contains an identical number of protons and electrons
   c) attracts other atoms
   d) repels other atoms
6. Protons and electrons
   a) are found in the atomic nucleus
   b) have permanent electric charges of opposite character
   c) are the only elementary particles
   d) repel each other
7. The coulomb, international unit of charge,
   a) measures the flow of dynamic electricity
   b) is equal to 6.25 billion billion electrons in motion
   c) represents 6.25 billion billion electrons
   d) is equal to the combined charge of 6.25 billion billion electrons
8. Induced charges
   a) leave a body in a neutral state
   b) are due to lost or gained electrons
   c) are produced by friction
   d) can exist only in metals
9. Voltage is the name given to
   a) the electric charges
   b) dynamic electricity
   c) the electromotive force created by electric charges
   d) the lack or excess of electrons on a body
10. Voltage
    a) does not affect electrons
    b) is created only by dry cells
    c) cannot be measured directly
    d) tends to move electrons
11. Electron current
    a) always flows from a negative toward a positive charge
    b) can flow through any material
    c) is measured in coulombs
    d) can flow without a voltage acting on the free electrons
12. The ampere
    a) is the unit of electromotive force
    b) is a flow of 6.25 billion billion free electrons per second
    c) is equal to 6.25 billion billion electrons
    d) does not depend on time

13. The main purpose of any electric circuit is
    a) to send a controllable flow of energy to a load
    b) to convert static to dynamic electricity
    c) to provide a non-metallic pathway for electrons
    d) to convert kinetic to potential electric energy
14. Inside the load, free electrons
    a) encounter little opposition
    b) stop their flow
    c) gain kinetic energy
    d) are put to work
15. Resistance
    a) is encountered only in the load of a circuit
    b) converts potential to kinetic energy
    c) is caused by electron-atom collisions in conductors
    d) does not depend on the type of metal
16. In an electric circuit,
    a) all parts have the same resistance
    b) the load always has the highest resistance
    c) resistance does not affect electrons
    d) all resistance is concentrated in the load
17. The term conventional current
    a) was first used by William Gilbert
    b) is used in working with electron flow
    c) can be used in place of electron current
    d) refers to an assumed direction of charge flow

4

METHODS OF GENERATING ELECTRICITY

All sources of electricity are **energy converters**. In some, mechanical energy is changed to electricity. Others release the energy stored in chemicals. A third group converts heat and light energy to electricity.

4-1 MECHANICAL METHODS OF GENERATING ELECTRICITY

Sliding friction between non-metallic materials is the most common method of generating **static electricity.**

If a plastic rod is rubbed with fur, electric charges will be produced on the rod and on the fur. The rod will receive a negative charge, the fur, a positive one. The charge on the fur leaks away immediately. But the charge trapped on the rod has **potential electric energy**; it can be put to work. Most of the mechanical energy used in the charging process is **lost as heat.**

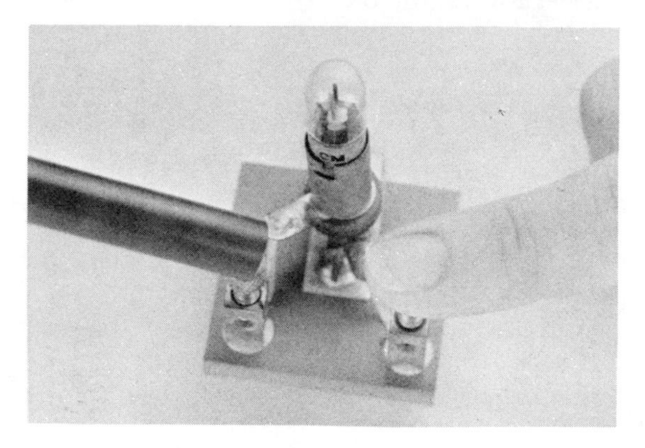

Figure 4-1 Discharging a Plastic Rod

That the charge on the rod really has potential electric energy can be proved quite easily. Touch one terminal of a small neon lamp with the charged end of the rod. The voltage of the rod's charge will drive a brief burst of electricity through the lamp. If the room is not too bright, you will see a flash of red light. The rod will have been discharged.

Only two generators of static electricity are in common use: the **Wimshurst machine** and the **Van de Graaff generator**.

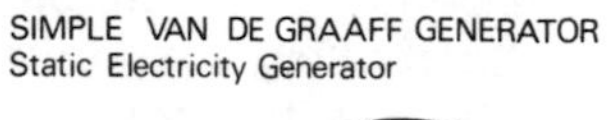

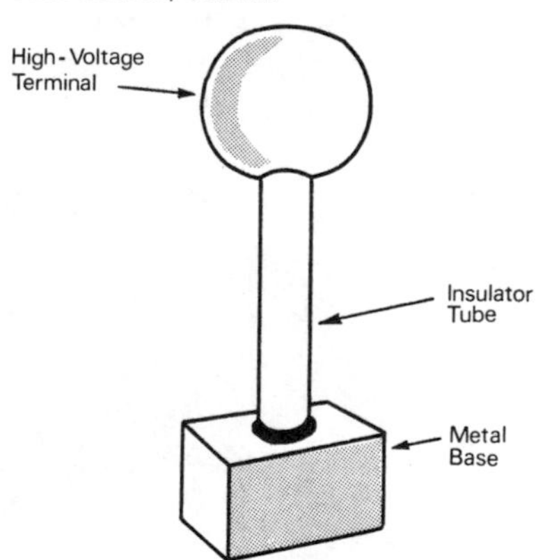

Figure 4-2 Van de Graaff Generator

Static electricity is often the undesired by product of friction between moving objects. Clouds driven by strong winds can gather huge electrostatic charges. When they discharge to the earth, the resulting lightning bolt often tears apart everything in its path.

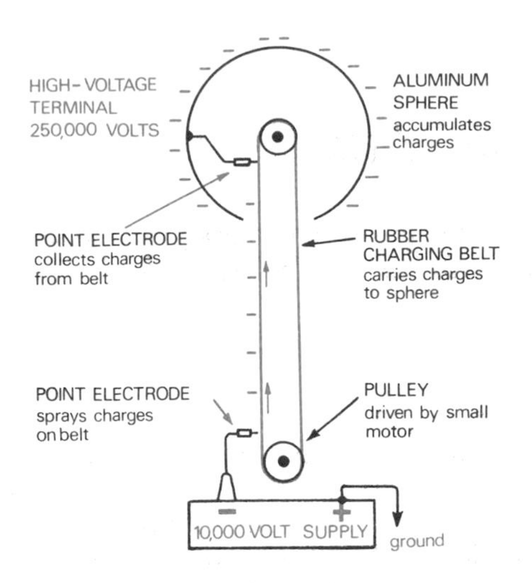

Figure 4-3 Main Parts of Van de Graaff Generator —Simplified

A moving magnetic force cutting through copper conductors is the most common method of generating **dynamic electricity**.

If you move a strong bar magnet up and down in the centre of a coil, as shown in Figure 4-4, the **moving** magnetic force will **induce a voltage** in the coil. This voltage will drive an electron current through the coil-meter circuit.

Figure 4-4 Inducing a Voltage in a Coil

Inside the copper conductors the free electrons are either pushed or pulled into motion by the moving magnetic force if the circuit is closed. As you move the magnet up and down, you **reverse** the direction of its motion. In the same manner, the free electrons reverse their direction inside the copper conductors. This back-and-forth flow is called **alternating current or a.c.**

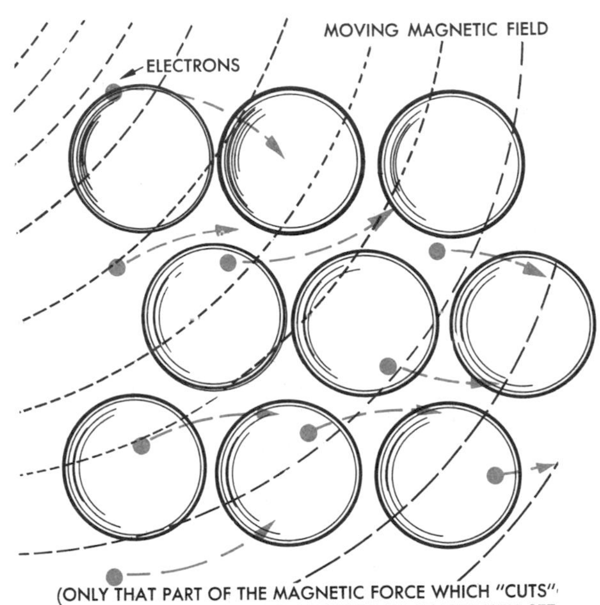

Figure 4-5 Action of Moving Magnetic Force on Free Electrons

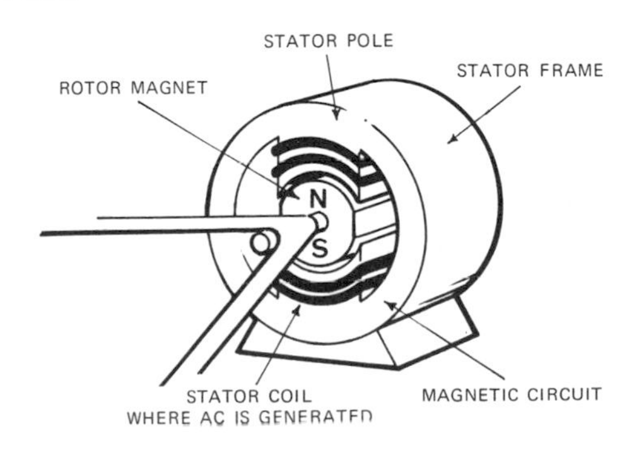

Figure 4-6 Simple Alternator

This method of generating electricity is called **electromagnetic induction**. It is highly efficient. The gigantic electric alternators in power plants around the world work on this principle. (Figure 4-6 shows the main parts of a simple alternator.)

Changing the pressure on a crystal is a third method of generating electricity by mechanical means. Certain crystalline substances such as **quartz, Rochelle salt** and **lead zirconium titanate** generate an alternating voltage when subjected to a changing pressure. This voltage appears on the metallized surfaces of the crystal. Its **polarity changes** as the pressure on the surfaces is changed.

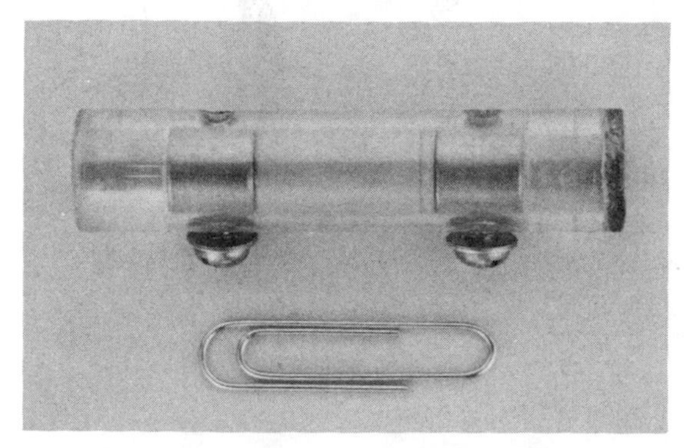

Figure 4-7 A Piezoelectric Crystal

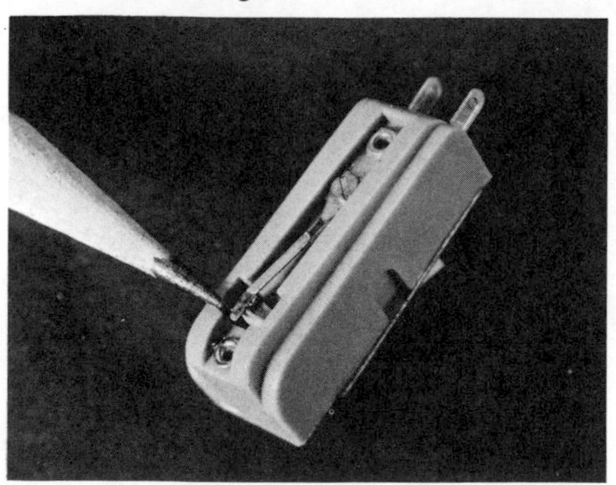

Figure 4-8 Crystal Cartridge for Record Player

If a small neon lamp is connected to a piezoelectric crystal, and if the crystal is hit with a rubber mallet, a brief flash of light will occur. A tiny portion of the mechanical energy will have been converted to electricity.

Figure 4-9 Crystal Microphone (Diaphragm partially removed)

Piezoelectric crystals are used mostly in phonograph pickups and in microphones, where they convert mechanical vibrations into weak electrical signals.

4-2 CHEMICAL METHODS OF GENERATING ELECTRICITY

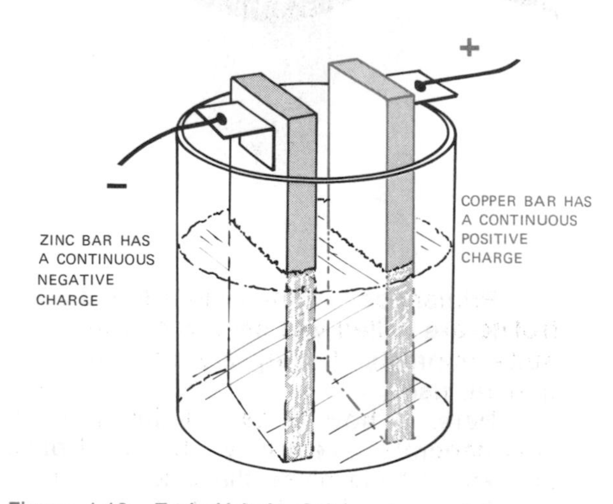

Figure 4-10 Early Voltaic Cell

In 1800 A.D., Alessandro Volta invented the first generator of dynamic electricity, now called the **voltaic cell**. Volta put two strips of **unlike metals**, copper and zinc, into a glass jar filled with **brine**. The **chemical reaction** between the brine and the metals produced the first continuous flow of electricity.

Chemical generators which are gradually used up as they continue to produce electricity are now called **primary cells**. A newer type of cell that can be recharged many times is the **secondary cell**. Both types use Volta's original method of immersing two unlike metals in an acid solution.

4-2A The Primary Cell

The metal rods or strips in a primary cell are called the **electrodes**, and the acid solution, the **electrolyte**. Various combinations of metals and acids are possible, and each will generate a voltage and polarity characteristic for that combination.

Figure 4-11 Modern Wet Cell

Primary cells containing a **liquid electrolyte** are called **wet cells**. Since they are quite impractical, they are no longer in general use.

Here is how a carbon-zinc primary cell generates electricity: the electrolyte attacks and dissolves the zinc electrode, but does not affect the carbon. Each zinc atom torn away by the acid gives up two free electrons to the zinc, thus generating a **negative charge** on this electrode.

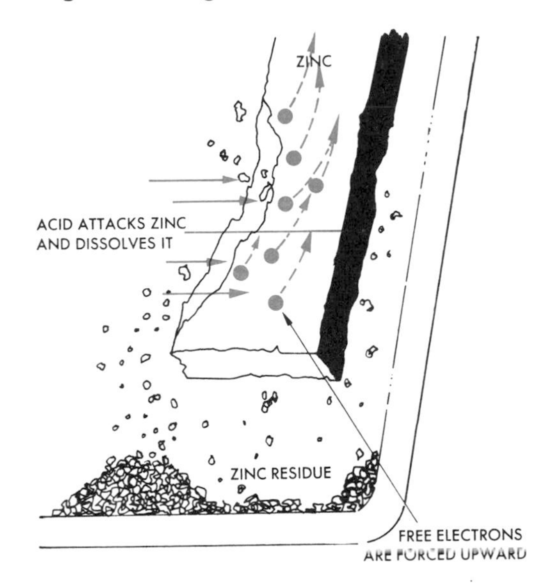

Figure 4-12 Chemical Action Inside a Primary Cell

At the same time, the electrolyte pulls free electrons out of the carbon rod, giving that electrode a **positive charge**. The resulting voltage will drive an electron current through a load as long as metallic

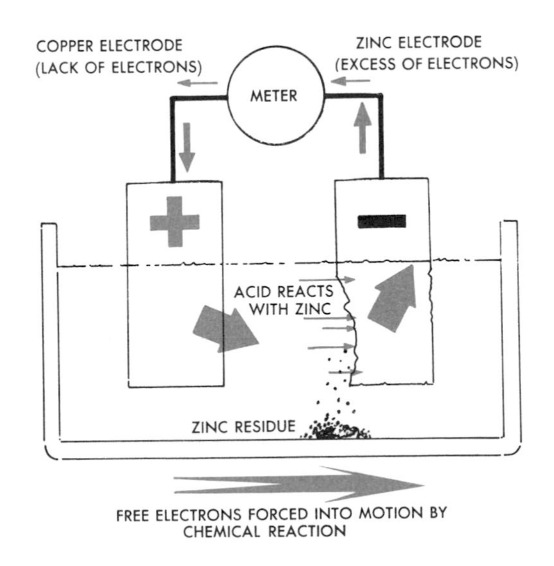

Figure 4-13 Experimental Carbon-Zinc Cell at Work

zinc is left in the cell. The characteristic voltage produced by a **zinc-carbon cell** is about **1.5 volts**.

4-2B The Secondary Cell

A secondary cell converts chemical to electric energy in the same way as a primary cell. Unlike a primary cell, however, it must first be **charged** before it can deliver electricity, and it can be **recharged** hundreds of times when run down.

The most common secondary cell is the lead-acid cell. In its simplest form, its electrodes are pure lead, and its electrolyte is dilute sulphuric acid. When new, such a cell will not generate electricity, because both electrodes are of the same metal. The cell can be charged by connecting it to a low-voltage D.C. power supply.

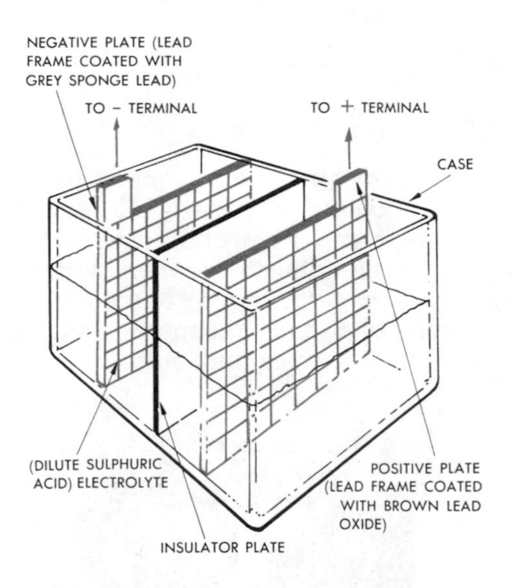

Figure 4-14 Simple Lead-Acid Secondary Cell

During the **charging** process a brownish substance, lead dioxide, is formed on one lead strip while new lead is deposited on the other. Thus, **the charging current changes the like metals** into unlike substances, creating a voltaic cell. Also, the acid content of the electrolyte increases as hydrogen gas bubbles out of it.

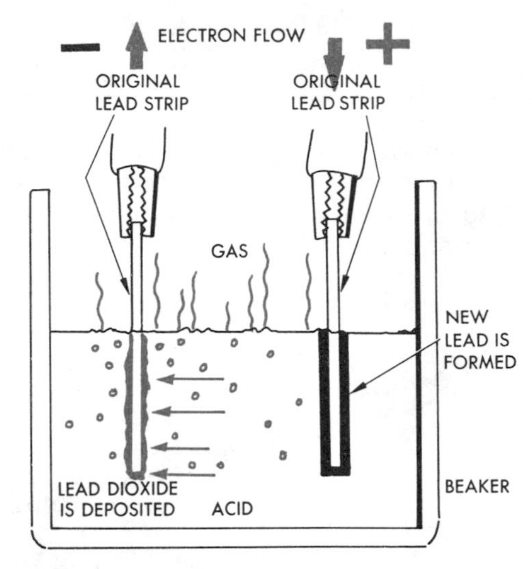

Figure 4-15 Charging the Simple Lead-Acid Cell

When the secondary cell is **discharged**, the acid attacks the lead electrode and wears away the lead dioxide on the other electrode, generating a voltage of about **2.2 volts**. Also, the acid content of the electrolyte decreases until it approaches water in composition.

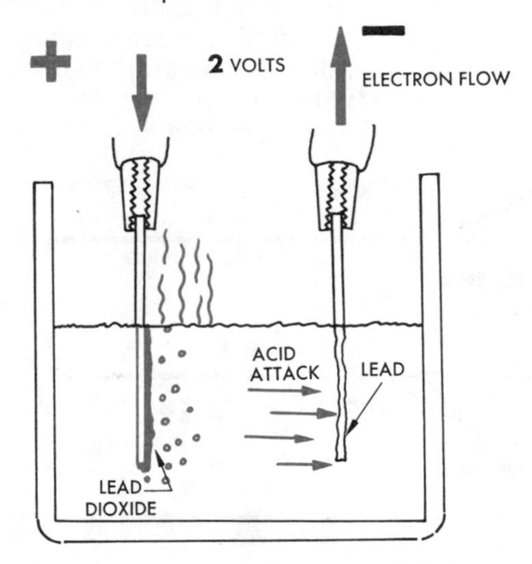

Figure 4-16 Discharging the Simple Lead-Acid Cell

Caution: the hydrogen gas produced during the charging of a secondary cell forms an explosive mixture with the surrounding air.

4-3 CONVERTING HEAT AND LIGHT ENERGY TO ELECTRICITY

Two methods of changing heat and light energy to electricity are in common use: heating a thermocouple, and shining light on a solar cell.

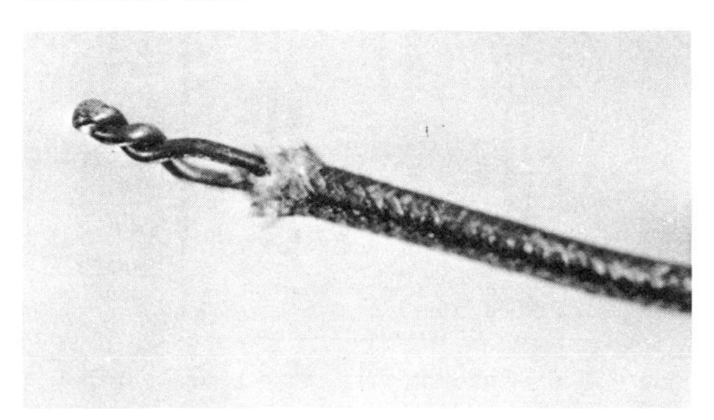

Figure 4-17 A Thermocouple

4-3A The Thermocouple

If the **welded junction** of two unlike metals such as nickel and brass **is heated**, the heat energy forces the free electrons in the nickel to flow into the brass, across the junction. The resulting electric charges create a voltage of several millivolts (1/1000 of a volt) as long as the heat energy flows into the metals.

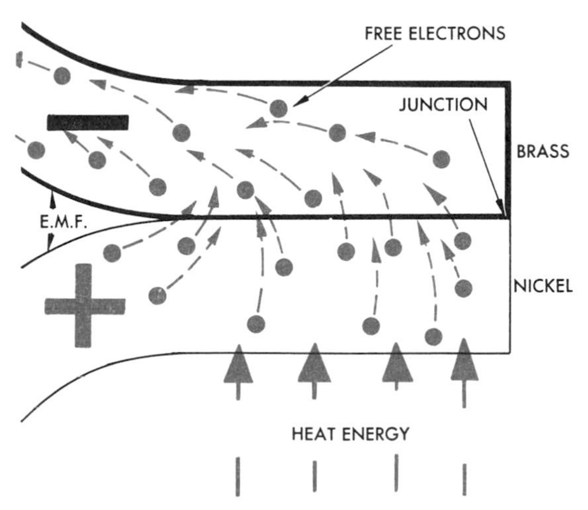

Figure 4-18 Electron Flow in a Heated Thermocouple

Thermocouples are extremely inefficient. To obtain a useful output, many of them must be connected together to form a **thermopile**. Their combined output is sufficient to operate other electrical devices such as safety valves, flame detectors, and thermometers.

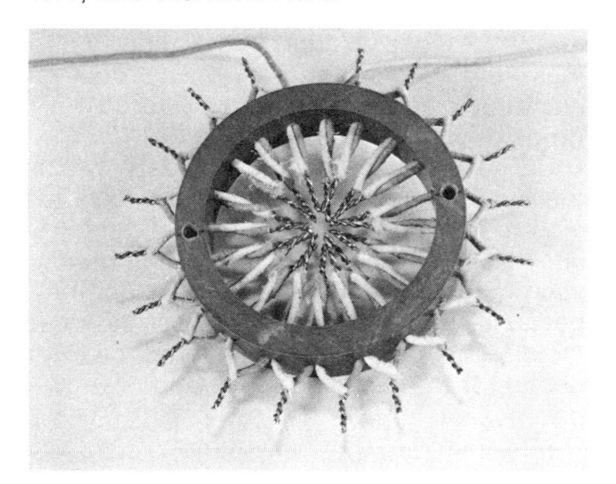

Figure 4-19 Thermopile

4-3B The Solar Cell

If light energy strikes the **junction** between two semiconductor materials, free electrons are forced from one semiconductor layer into the other. The semiconductor material facing the light source must be extremely thin and transparent to allow the light to reach the junction.

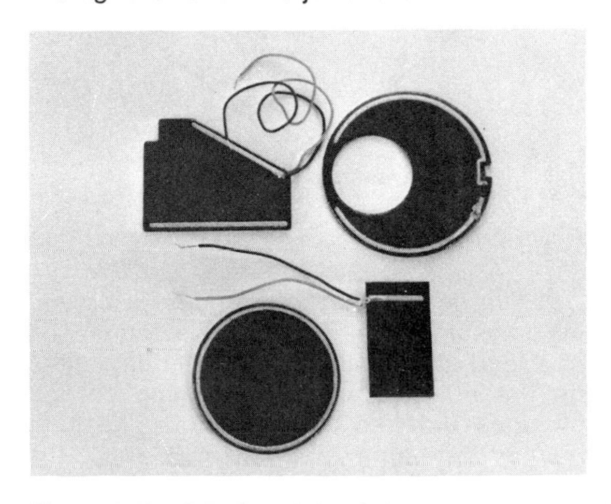

Figure 4-20 Selenium Solar Cells

Figure 4-20 shows several selenium solar cells. Other, more efficient cells are made of **germanium** or **silicon**. All work in a similar way, generating voltages up to 0.4 volts when light is shining on them.

The most efficient solar cell is the silicon cell. It converts up to 14 percent of the light energy to electricity. (The remainder is lost as heat.) Under normal operating conditions, its life span is practically unlimited.

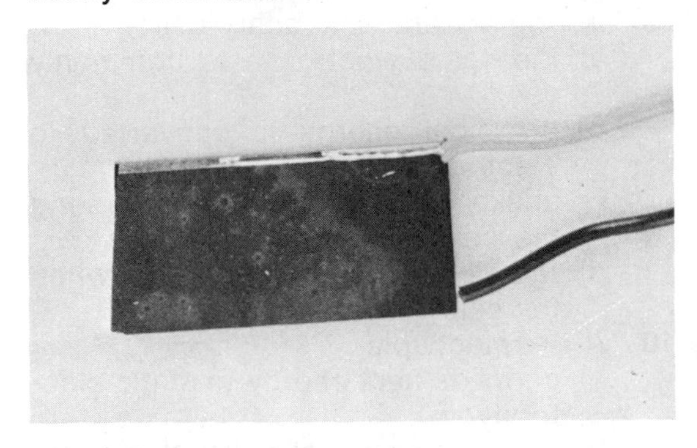

Figure 4-21 Silicon Solar Cell

4-4 ASSIGNMENTS: METHODS OF GENERATING ELECTRICITY

4-4A Write Full Answers

1. Explain what all sources of electricity have in common.
2. Describe a simple way to prove that a charged plastic rod has potential electric energy.
3. Name two generators of static electricity that are in common use.
4. State the most common method of generating dynamic electricity.
5. Explain in detail how electricity is generated by electromagnetic induction.
6. What is a piezoelectric crystal?
7. List several important uses for piezoelectric crystals.
8. Draw a sketch of Volta's original cell and explain how it works.
9. Explain the differences between a primary and a secondary cell.
10. Why is it dangerous to charge a secondary cell in an unventilated room?
11. Describe with the aid of a drawing how a thermocouple generates dynamic electricity.
12. Find out how a thermocouple is used as a safety device in a gas furnace.
13. Make a simple sketch of the cross-section of a solar cell and explain how the cell works.

4-4B Indicate True or False

1. The fur used in charging a plastic rod does not receive an electric charge.
2. Static electricity is often the undesired by-product of friction between moving objects.
3. The charge on a plastic rod rubbed with fur exerts a force on free electrons.
4. When a moving magnetic force cuts through an insulator, an electron current will be generated.
5. Alternating current is generated by electromagnetic induction.
6. A moving magnetic force generates static electricity in a conductor.
7. A piezoelectric crystal converts a steady pressure to an alternating voltage.
8. Some important piezoelectric materials are Rochelle salt, quartz, and lead zirconium titanate.
9. A piezoelectric crystal produces static electricity when hit with a mallet.
10. A voltaic cell generates dynamic electricity by means of a chemical reaction.
11. A primary cell consists of two like metals and an electrolyte.
12. A discharged secondary cell is of no further use.
13. The characteristic output voltage of a carbon-zinc primary cell is 1.5 volts.
14. A thermocouple efficiently converts heat to electric energy.
15. A thermocouple generates a few millivolts when heated with a flame.
16. A silicon solar cell converts up to 14 percent of the light energy to electricity.

4-4C Select Correct Answer

1. When a plastic rod is charged with fur,
   a) the fur receives a negative charge
   b) the rod loses electrons
   c) both rod and fur acquire an electric charge
   d) the rod loses its charge immediately
2. If a charged plastic rod touches the terminals of a neon lamp,
   a) the lamp also becomes charged
   b) the charge on the rod remains unchanged
   c) a brief flow of electrons will flash the lamp
   d) no energy conversion takes place
3. Dynamic electricity can be generated in a conductor
   a) by a moving magnetic force
   b) by electrostatic induction
   c) by a stationary magnetic force
   d) by friction
4. Electromagnetic induction
   a) generates static electricity
   b) is a method of producing electrons
   c) occurs in all non-metallic materials
   d) generates alternating current
5. A moving magnetic force cutting through a conductor
   a) vibrates the metal atoms
   b) pushes and pulls the free electrons into motion
   c) has no effect on free electrons
   d) generates static electricity
6. A piezoelectric crystal
   a) converts constant pressure to electricity
   b) generates static electricity when subjected to pressure
   c) generates an electron flow at all times
   d) converts a changing pressure to an alternating voltage
7. Piezoelectric crystals
   a) convert mechanical vibrations to electrical signals
   b) are used to generate large amounts of electricity
   c) can be made with any non-metallic material
   d) generate electricity continuously
8. Both primary and secondary cells
   a) can be recharged many times
   b) produce static electricity
   c) use two unlike metals and an acid to generate electricity
   d) were invented by Alessandro Volta
9. In primary and secondary cells,
   a) the same electrolyte is commonly used
   b) chemical energy is converted to electricity
   c) unlike metals are used in making the cells
   d) the electrolyte is not affected when the cell is discharged
10. A thermocouple
    a) converts heat energy to static electricity
    b) generates several volts of electromotive force
    c) converts heat energy to dynamic electricity
    d) produces alternating current when heated
11. A solar cell
    a) converts light energy to dynamic electricity
    b) generates static electricity when illuminated
    c) can be made of any non-metallic material
    d) consists of two thick layers of semiconductor material
12. A modern silicon solar cell
    a) generates an output voltage of 0.04 volts in full sunlight
    b) converts all the incident light energy to electricity
    c) converts up to 14 percent of the incident light energy to electricity
    d) is made of two layers of selenium

5

PRACTICAL SOURCES OF ELECTRICITY

Only a few of the many methods of generating electricity are put to large-scale use. Most of the electric energy used in homes and industry is generated by huge alternators driven by steam or water turbines. Portable and mobile equipment is usually operated with dry cells or storage batteries. Spacecraft receive their electric energy from large panels of solar cells.

5-1 D.C. AND A.C. SOURCES OF ELECTRICITY

Practical generators of electricity fall into two general groups: **direct current** or **d.c.** sources, and **alternating current** or **a.c.** sources.

Direct current or d.c. is produced by those generators that do not change the polarity of their terminals. To this group belong all cells, batteries, thermocouples, solar cells, and rotating d.c. generators.

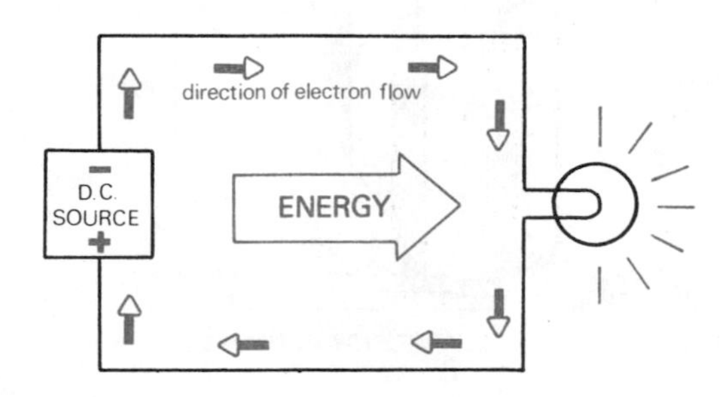

Figure 5-1 Energy Transfer with D.C.

If free electrons are driven through a circuit by a d.c. source, they never change their direction of flow, producing a direct current, or d.c.

An **a.c. source** continuously reverses the polarity of its terminals. This forces the free electrons in a circuit to flow back and forth, producing a continuously alternating current, or a.c. Both d.c. and a.c. can transport electric energy to a load with equal efficiency.

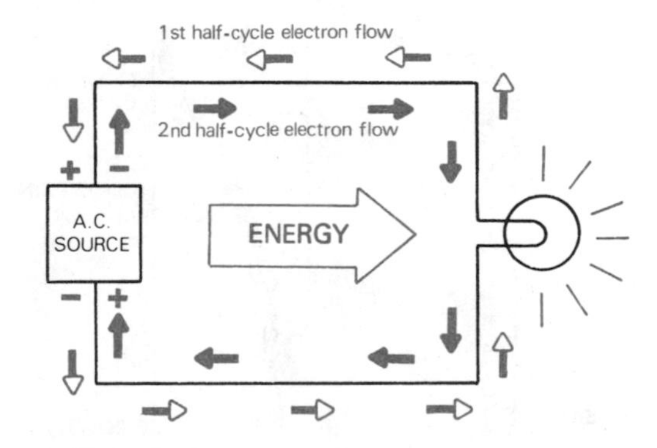

Figure 5-2 Energy Transfer with A.C.

A.c. is commonly generated by electromagnetic induction in huge alternators (a.c. generators). Its voltage and current levels can be changed easily in **transformers**.

5-2 DRY CELLS

Dry cells are primary cells whose electrolyte is mixed with starch to form a thick paste. The cell is sealed and enclosed in a steel container to make it a portable and safe source of electricity.

Industry has developed three general types of dry cells: the **carbon-zinc** cell, the **manganese alkaline** cell, and the **mercury** cell. Each of these cells uses a different metal-electrolyte combination, and each has specific advantages over the others.

5-2A The Carbon-Zinc Dry Cell

This is the most common dry cell in general use. It is made in a great variety of sizes and shapes, as shown in Figures 5-7 and 5-8. In the carbon-zinc cell, electricity is generated by the chemical reaction between **ammonium chloride** (sal ammoniac) and **zinc**. All other materials in the cell support this basic reaction.

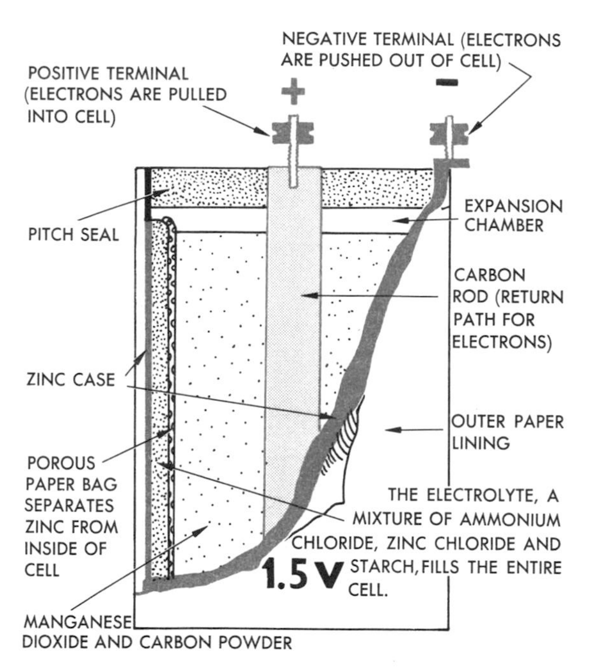

Figure 5-3 Internal Construction of a Carbon-Zinc Dry Cell

The ammonium chloride electrolyte attacks the zinc and gradually dissolves it, generating a voltage of about **1.5 volts**, regardless of cell size. The zinc casing is the negative, and the carbon rod, the positive terminal of the cell. A manganese-carbon mixture, the **depolarizer**, surrounds the central carbon rod; it prevents the formation of hydrogen gas inside the cell.

Carbon-zinc cells have only a limited shelf life. Undesirable chemical reactions, called **local action**, eventually destroy the zinc electrode even when the cell is not in use. For this reason, carbon-zinc cells must be **removed** from equipment that is to be stored for long periods. Once local action penetrates the cell casing, the equipment itself will be attacked by the electrolyte.

5-2B The Manganese Alkaline Dry Cell

A manganese alkaline cell can supply much more energy than a carbon-zinc cell of the same size. Alkaline cells are made in the same sizes and shapes as the carbon-zinc types and can be used as direct replacements for them.

The inner cylinder of finely divided **zinc** mixed with mercury forms the nega-

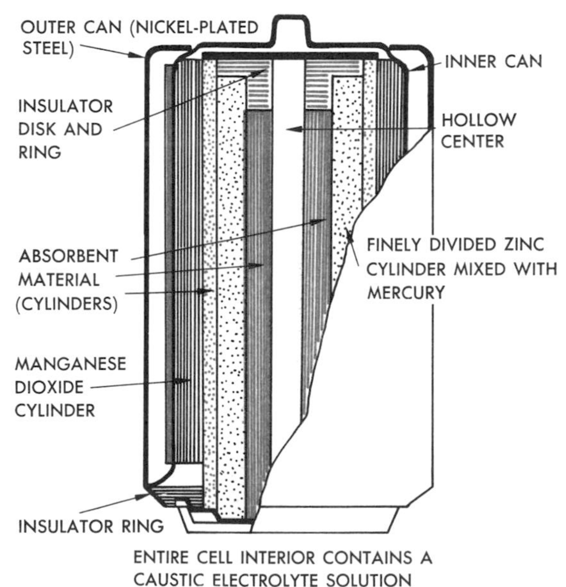

Figure 5-4 Internal Construction of a Manganese Alkaline Cell, Size D

tive electrode of the cell, and the outer cylinder of **manganese dioxide** forms the positive electrode. The chemical reaction between the alkaline electrolyte paste and the zinc generates approximately **1.5 volts** when the cell is new. The cell is tightly sealed.

5-2C The Mercury Cell

The mercury cell has several important advantages over the carbon-zinc and the manganese alkaline cells: it can be stored for long periods of time without losing its charge, has a very high energy output, maintains an almost constant output voltage and can supply heavy bursts of energy without internal damage.

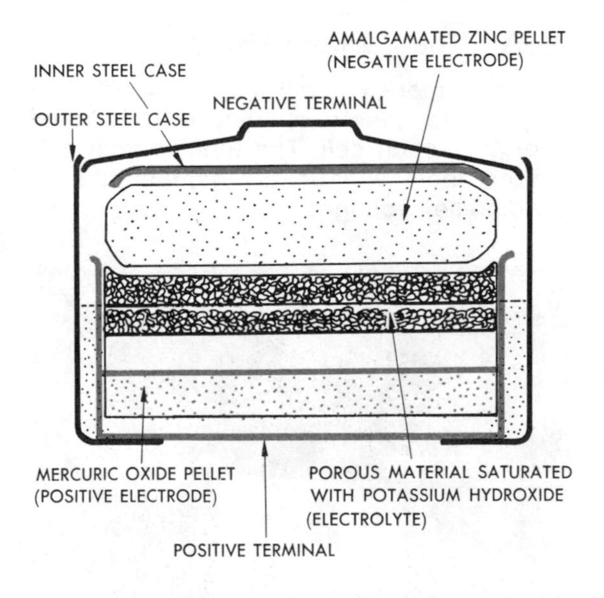

Figure 5-5 Internal Construction of a Typical Mercury Cell

In the cell shown in Figure 5-5, the negative electrode is a **zinc-mercury** mixture, and the positive electrode, a pellet of **mercury oxide**. The electrolyte, **potassium hydroxide**, is held in the pores of a filler material. As in all dry cells, the principal chemical reaction takes place at the zinc electrode, generating a nearly constant output of **1.35 volts**.

The mercury cell is the most expensive of the common dry cells. Its reliability makes it the preferred energy source for electronic watches and other precision instruments.

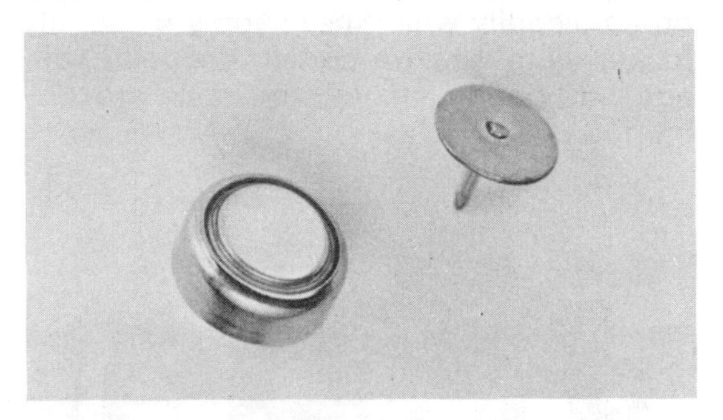

Figure 5-6 Mercury Button Cell

5-2D Dry Cell Data

All types of dry cells are made in a great variety of sizes and shapes. The most popular sizes, called the AA, the C, the D and the Number 6 cell, are shown in Figures 5-7 and 5-8.

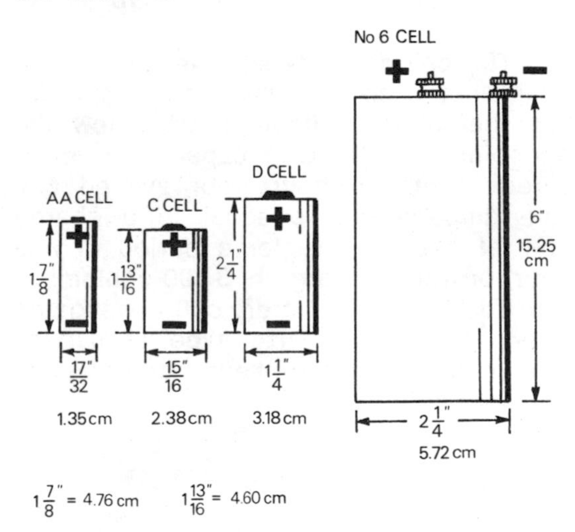

Figure 5-7 Physical Dimensions of Popular Cell Sizes

Two electrical ratings are important in working with dry cells: the cell voltage and the cell capacity. **The cell voltage**, which is completely independent of cell size, is determined by the cell materials.

The typical output voltage of a fresh cell is 1.55 volts for zinc-carbon cells, 1.55 volts for manganese alkaline cells, and 1.35 volts for mercury cells.

The output voltage of all dry cells drops steadily when they supply electricity to a circuit. When a certain **endpoint voltage** has been reached, the cells are discarded.

Figure 5-8 Popular Dry Cells, Sizes C, D, and AA

The cell capacity is a measure of the total electric charge which a cell can deliver before its voltage drops below the endpoint voltage. Cell capacity is measured in **ampere-hours**, abbreviated **a.h.** One ampere-hour is equal to a charge flow of one ampere for a period of one hour, or a total charge of 3,600 coulombs.

The capacity of a dry cell depends on three factors: **cell size, type of service** (continuous or intermittent), and desired **endpoint voltage**.

CELL CAPACITY IN AMPERE-HOURS

CELL SIZE	CARBON-ZINC CELL	ALKALINE CELL	MERCURY CELL
AA	0.125	0.425	2.4
C	0.8	3.1	–
D	2.0	8.5	–
6	25.0	–	–

Table 5-1 Capacity Ratings of Important Dry Cells

The schematic symbol for a dry cell is shown in Figure 5-9.

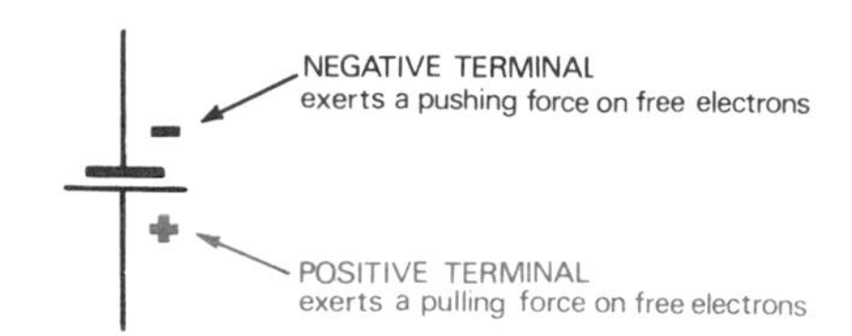

Figure 5-9 Schematic Symbol for a Dry Cell

5-3 BATTERIES

Dry cells may be connected in **series** to form a battery with a **higher voltage** than a single cell, or in **parallel** to make a battery with a **larger capacity**, or total charge.

5-3A The Series Battery

If two or more dry cells are connected in series, the electrons receive **additional energy** in each cell. The **total voltage** acting on them will be the **sum** of the individual cell voltages.

Figure 5-10 A Series Battery

Three Number 6 dry cells in series are shown in Figure 5-10. If the cells are tresh, the total voltage of the series battery will be 4.5 volts, but its capacity will only be that of a single cell, 25 ampere-hours.

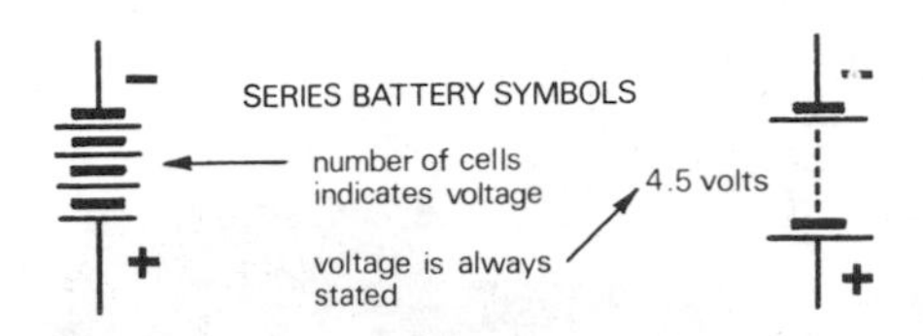

Figure 5-11 Series Battery—Schematic Symbols

Since all cells in a series battery are **discharged simultaneously**, they are all used up together. The two schematic symbols for a series battery are shown in Figure 5-11.

Practical series batteries are made in many sizes, shapes and output voltages. In some batteries, especially the 9-volt transistor battery, the individual dry cells are flat wafers pressed together tightly and connected with a special conducting compound, silver wax.

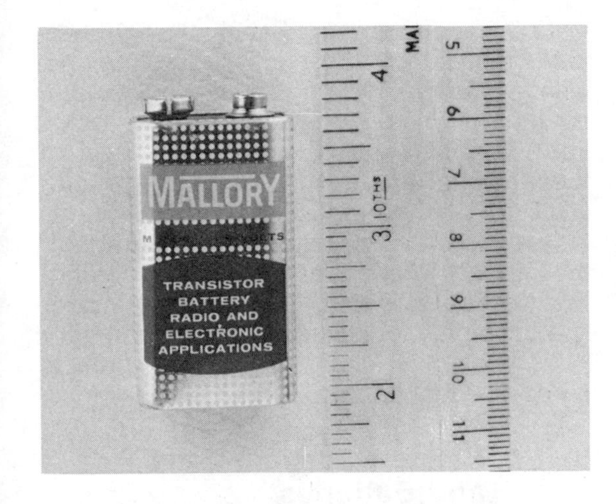

Figure 5-12 9-Volt Transistor Battery

5-3B The Parallel Battery

When a battery is needed with a **greater capacity** than a single cell, two or more dry cells are connected in **parallel**, permitting each cell to supply a part of the total electron current flowing through the load.

Figure 5-13 A Parallel Battery

Cells are connected in parallel by linking their negative and positive terminals with wires, as shown in Figure 5-13. Do not cause an accidental **short circuit** by connecting a wire to the terminals of the same cell, it will ruin the cell.

The output voltage of a parallel battery is the same as that of a single cell, regardless of the number of dry cells used. The **total** capacity of the battery equals the sum of the cell capacities. Thus, three Number 6 dry cells in parallel have a voltage of 1.5 volts and a total capacity of 75 ampere-hours.

Few parallel-cell batteries are made by manufacturers. They prefer to make a bigger dry cell if more capacity is required.

5-4 STORAGE BATTERIES

A storage battery consists of several **secondary cells** connected in **series.** Two kinds of secondary cells are widely used for this purpose: the **lead-acid cell** and the **nickel-cadmium cell**. Both types have distinct advantages and drawbacks. The lead-acid cell is the most common, especially in the automotive industry.

5-4A The Lead-Acid Storage Battery

The common lead-acid storage battery consists of three or six lead-acid secondary cells. These cells can deliver large amounts of electron current for short periods, as required for starting an automobile engine.

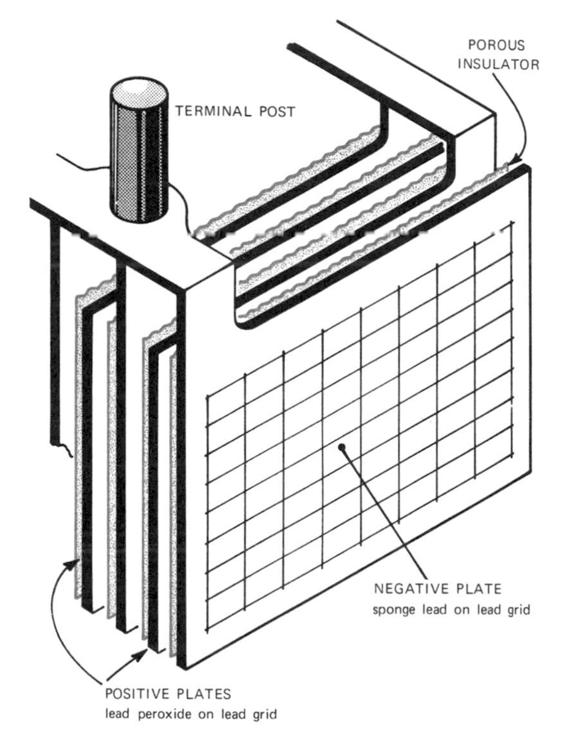

Figure 5-14 A Lead-Acid Secondary Cell—Simplified

A typical lead-acid secondary cell has a negative electrode made of **sponge lead**, and a positive electrode made of **lead dioxide**. Its electrolyte is **dilute sulphuric acid**.

Although only one pair of electrodes is shown in the simplified drawing of Figure 5-14, a standard lead-acid cell has eight or ten such electrode pairs to increase cell capacity. Its output voltage ranges from **2.0 volts** when nearly discharged to **2.2 volts** when fully charged.

Figure 5-15 Typical Lead-Acid Battery

The common lead-acid storage battery comes in 6-volt and 12-volt versions, with capacities ranging from about 30 a.h. for motorcycles to over 200 a.h. for trucks. Four problem areas restrict the use of these batteries to the automotive field: (1) During charging, hydrogen gas is released, forming an explosive mixture with the air. (2) The battery must be operated in a vertical position. (3) Its liquid electrolyte is highly corrosive. (4) If the battery is too quickly charged its life span is greatly reduced.

5-4B Nickel-Cadmium Cells and Batteries

Nickel-cadmium secondary cells are much more versatile than lead-acid cells and much more expensive. They are available both as single cells in the most popular cell sizes and as batteries.

The principal materials used in this type of cell are **cadmium** (negative electrode), **nickel oxide** (positive electrode), and **potassium hydroxide** paste (electrolyte). Typical output voltage is **1.25 volts to 1.35 volts**, and the cell should be recharged when its voltage drops below 1.1 volts.

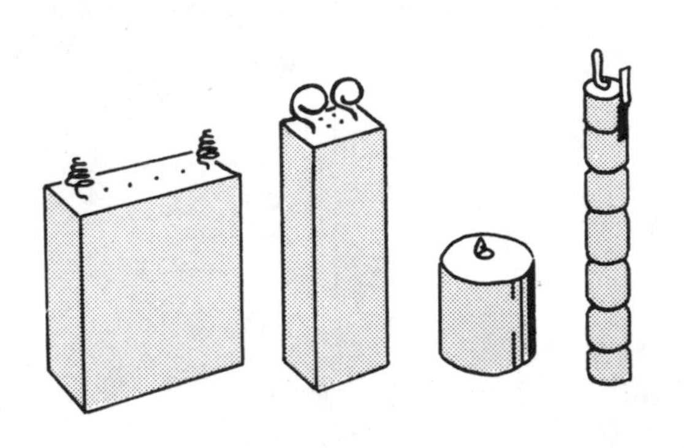

Figure 5-16 Common Nickel-Cadmium Cells

Nickel-cadmium cells are completely sealed and can be operated in any position. Their extreme ruggedness and reliability have spurred the development of new devices such as battery-operated TV sets and power tools.

5-5 THE TRANSFORMER

A transformer can change, or **transform**, both voltage and current levels of an **alternating current** to any desired value. It is strictly an **alternating current device** and will **burn up** if accidentally connected to a d.c. source.

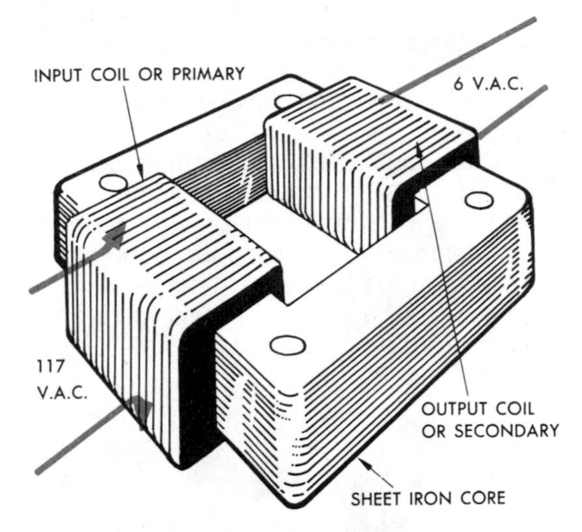

Figure 5-17 Simple Transformer

A transformer has three basic parts, as shown in Figure 5-17: a **primary** or input coil, a **secondary** or output coil, and a **laminated iron core**.

Alternating current is fed into the primary coil and **converted** to continuously expanding and contracting lines of magnetic force. This magnetic form of energy is conducted to the secondary coil in the laminated iron core. In the windings of

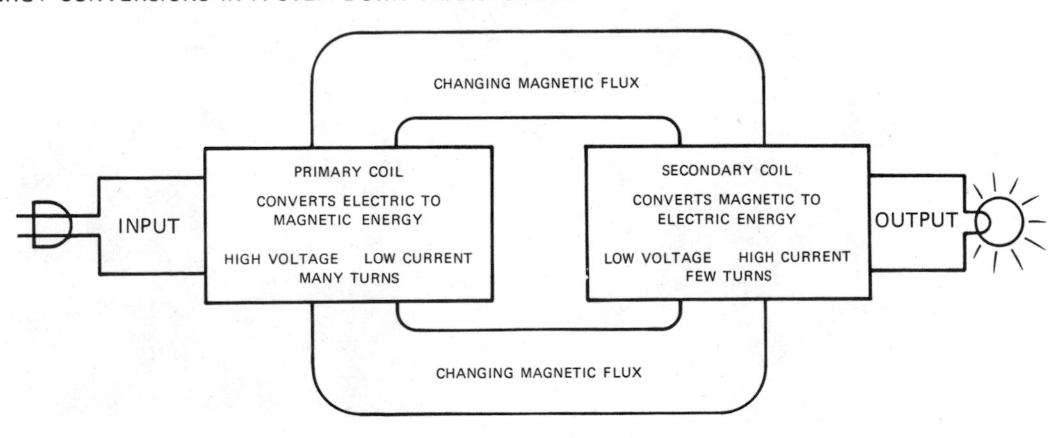

Figure 5-18 Energy Conversions in a Transformer

the secondary coil, this changing magnetic force **induces** new alternating current by electromagnetic induction, at a voltage and current level different from those in the primary. The exact value of the secondary voltage depends on the **number of turns** in the two coils and on the primary voltage.

A well-designed transformer is highly efficient, since little energy is lost in this double conversion. Some loss occurs in the resistance of the copper conductors, and some in the iron core. The greater this energy loss, the more heat will develop in the transformer.

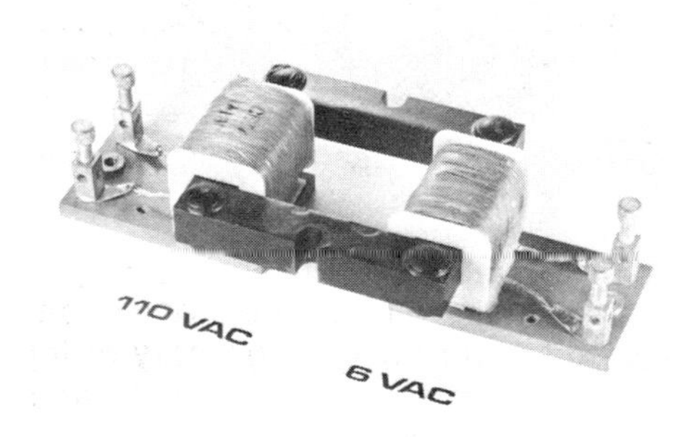

Figure 5-19 Simple Transformer, 110VAC to 6VAC

In many transformers, the two coils are wound on top of each other, as in the bell transformer in Figure 5-20. Complex transformers can have several secondary coils to provide different output voltages.

Figure 5-20 Bell Transformer, Cover Removed

Since there is **no electrical connection** between primary and secondary coils, a transformer **isolates** the equipment it operates from the power line. This greatly reduces the danger of electric shock.

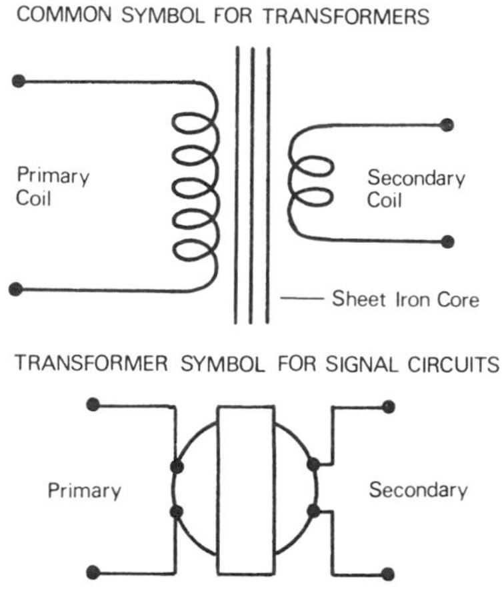

COMMON SYMBOL FOR TRANSFORMERS

Primary Coil

Secondary Coil

Sheet Iron Core

TRANSFORMER SYMBOL FOR SIGNAL CIRCUITS

Primary

Secondary

Figure 5-21 Schematic Symbols for Transformers

Figure 5-21 shows the **schematic symbols** for a common transformer. It is customary to indicate the primary and secondary voltages on schematic diagrams.

5-6 THE LOW-VOLTAGE D.C. POWER SUPPLY

This device **converts** the 115-volts a.c. line voltage to a low d.c. voltage, and **isolates** its d.c. output from the power line. A low-

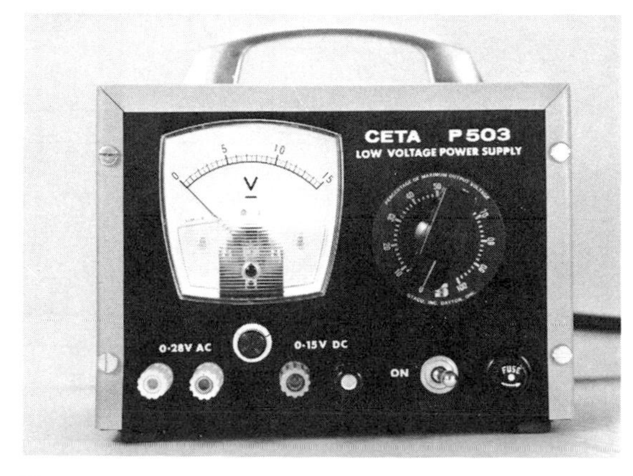

Figure 5-22 Low-Voltage Power Supply

voltage d.c. power supply can be used in place of a battery or to charge storage batteries.

The output voltage of most power supplies can be varied from 0 to about 15 or 20 volts d.c. But if no load is connected, this voltage may rise to more than 30 volts. A voltmeter continuously measures the output of the device.

When a low-voltage power supply is used with experimental circuits the following procedure will **safeguard** the instrument:

1. Compare the finished circuit with the schematic diagram **before** you connect it to the power supply.
2. Make sure that the main switch of the supply is in the **off** position and its voltage control at **zero** when you connect the circuit.
3. Have the completed hook-up **checked** by your teacher.
4. Turn the power supply on and rotate the voltage control **slowly** until the voltmeter indicates the desired voltage.
5. Watch the **ammeter** to prevent excess current levels in case of a circuit fault.

5-7 A.C. AND D.C. GENERATORS

Figure 5-23 A Typical Alternator

Probably more than ninety percent of all the electric energy used in North America is generated by rotating a.c. generators called **alternators**. These machines are usually driven by water or steam turbines. They generate a.c. by **electromagnetic induction**. Their rotating speed is the same, or **synchronized**, all over the continent, no matter where they may be located. This makes possible the pooling of their outputs into large power distribution systems, or grids.

The current produced by these alternators reverses its direction exactly **sixty times each second**. These continuous reversals of electron flow are called the **frequency** of the alternating current; it is measured in cycles per second or in **hertz**. One hertz, the international unit of frequency, is equal to one cycle per second.

Figure 5-24 D.C. Generator

D.c. generators are used mainly for special purposes, such as in welding, in charging a battery and in operating the electrical system of certain automobiles. They generate pulsating d.c. by **electromagnetic induction**, similar to the alternator. A special rotary switch, the **commutator**, changes the direction of the output current during every second half-cycle to maintain a direct current flow in the output circuit.

5-8 ASSIGNMENTS: PRACTICAL SOURCES OF ELECTRICITY

5-8A Write Full Answers

1. What do both a.c. and d.c. sources have in common?
2. Why is a.c. the preferred kind of electricity for general use?
3. Make a labelled drawing showing the internal design of a Number 6 carbon-zinc dry cell.
4. Explain briefly the purpose of each part or material in the carbon-zinc dry cell.
5. List the main ingredients of a manganese alkaline dry cell and describe how the cell works.
6. Make a chart showing the type of electrodes, electrolyte, typical output voltage and shelf life of the three main types of dry cell.
7. Define the unit of cell capacity.
8. Six size D carbon-zinc dry cells are connected in series. Draw the schematic diagram of this series battery and state its output voltage and capacity.
9. Make a simplified drawing of a lead-acid secondary cell. Label and identify all parts and materials.
10. Name the main parts and materials of the nickel-cadmium secondary cell and explain how the cell generates electricity.
11. Draw a top view of a simple transformer. Label and identify all parts.
12. Explain carefully how a transformer changes or transforms a.c., and what will happen if it is accidentally connected to a d.c. source.
13. State the five rules to be followed when connecting a low-voltage power supply to an experimental circuit.
14. Define carefully the term frequency and its international unit of measurement, the hertz.

5-8B Indicate True or False

1. Both d.c. and a.c. can deliver energy to a load with equal efficiency.
2. In a carbon-zinc dry cell, the ammonium chloride is the depolarizer.
3. The manganese alkaline dry cell generates as much electricity as the carbon-zinc cell.
4. The mercury cell generates a nearly constant output voltage of 1.35 volts and has an extremely long shelf life.
5. Both series and parallel batteries deliver a higher voltage than a single cell.
6. A storage battery consists of two or more secondary cells.
7. A lead-acid secondary cell produces an explosive hydrogen gas/air mixture when it is charged.
8. The nickel-cadmium cell is manufactured in the same sizes and shapes as the lead-acid secondary cell.
9. A transformer converts electric energy to magnetic energy, and then converts the magnetic energy to electric energy again with relatively little energy loss.
10. A low-voltage power supply produces an a.c. output voltage that can either charge or replace a battery.

5-8C Select Correct Answer

1. Both a.c. and d.c.
   a) are generated only by electromagnetic induction
   b) reverse electron flow at regular intervals
   c) are generated in huge alternators
   d) can transport energy equally well
2. The carbon-zinc dry cell
   a) eventually wears out because both its electrodes are dissolved by the electrolyte
   b) generates 1.5 volts when new, regardless of cell size
   c) generates hydrogen gas as well as electricity
   d) has identical electrodes
3. The manganese alkaline cell
   a) has a much longer shelf life than the carbon-zinc cell
   b) cannot be used as a direct replacement for a carbon-zinc cell
   c) has the same electrolyte as a carbon-zinc cell

d) is identical to a carbon-zinc cell in output voltage and cell capacity

4. The most reliable, efficient, and constant-voltage dry cell is
   a) the manganese alkaline cell
   b) the original voltaic cell
   c) the carbon-zinc dry cell
   d) the mercury cell
5. Three factors are essential in the selection of a dry cell:
   a) cell capacity, cell size, and electrolyte
   b) cell voltage, cell capacity, and type of electrolyte
   c) cell size, output voltage, and shelf life
   d) cell voltage, cell capacity, and shelf life
6. The lead-acid secondary cell
   a) has pure lead electrodes and a dilute sulphuric-acid electrolyte
   b) can supply large bursts of current for short periods of time
   c) is totally sealed
   d) can be operated in any position
7. The nickel-cadmium secondary cell
   a) can be operated in any position
   b) has a liquid electrolyte
   c) should be recharged when its voltage drops to 1.25 volts
   d) is available only in the form of single cells
8. The output voltage and current of a transformer depend on
   a) the size of the laminated iron core
   b) the number of turns in the primary winding
   c) the number of turns in both coils, and on the primary voltage
   d) on the resistance of its copper windings
9. A low-voltage power supply
   a) is insensitive to short circuits
   b) converts the a.c. line voltage to a variable, low-voltage direct current
   c) does not isolate its output from the line voltage
   d) can produce any desired output voltage
10. Both a.c. and d.c. generators
    a) employ rotating switches, or commutators
    b) are used for the large-scale production of electricity
    c) can be synchronized to allow pooling of their outputs
    d) operate on the principle of electromagnetic induction
11. The frequency of an alternator
    a) is the number of complete reversals of polarity per second
    b) is measured internationally in cycles per second
    c) cannot be sychronized with that of other alternators
    d) is independent of time
12. The international unit of measurement for frequency
    a) is the cycle per second
    b) is defined as the number of polarity reversals
    c) is the hertz, defined as one cycle per second
    d) is not part of the metric system, or S.I.

6

CONDUCTORS, SEMICONDUCTORS AND INSULATORS

All materials can be grouped into three broad categories: conductors of electricity, semiconductors and non-conductors, or insulators. The semiconductors form a distinctly separate group. There is no great difference in the characteristics of poor conductors and low-quality insulators.

6-1 CONDUCTOR AND RESISTOR MATERIALS

Only four metals are considered really good conductors of electricity: silver, copper, gold and aluminum.

Silver is seldom used because it blackens easily when exposed to certain gases. **Copper** is the most common conductor material, followed closely by **aluminum**. **Gold** is used for plating the contact surfaces of delicate electrical connectors because it is virtually corrosion proof.

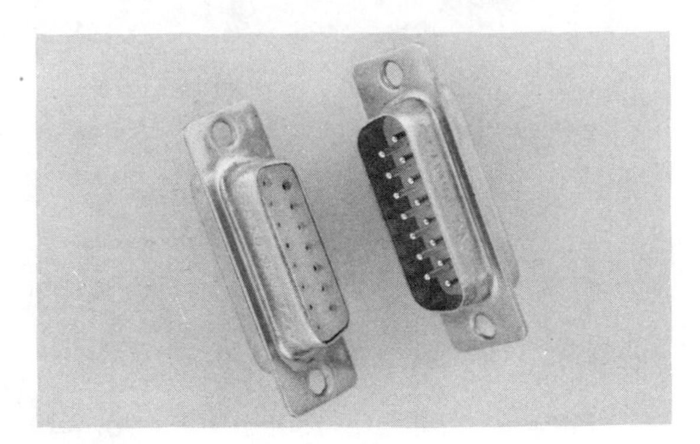

Figure 6-1 Gold-plated Computer Connector

Various metals and alloys differ greatly in their ability to conduct electricity. The common method of comparing them is to measure their **specific resistance**. For this purpose, the metals are drawn into thin, round wires, one foot long and one mil (1/1000 inch) in diameter (one circular mil-foot). The resistance of these wires is then measured and listed in special **resistivity tables**, as shown in Table 6-1. (In the metric system, the specific resistivity of a material is measured in ohms per metre.)

THE SPECIFIC RESISTANCE OF COMMON METALS AT ROOM TEMPERATURE
in ohms per circular mil–foot

SILVER	9.80 OHMS
COPPER	10.37 "
GOLD	14.70 "
ALUMINUM	17.02 "
TUNGSTEN	33.20 "
NICHROME	660.00 "

Table 6-1 The Specific Resistance of Common Metals

Constantan and **nichrome** are called **resistor** materials because their specific resistance is very high compared to that of the good conductors. These alloys are used for making **heating elements** and commercial **resistors.**

Many other metals are important to electrical technology, especially **molybdenum, palladium** and **platinum** for the contact pellets of switches; **mercury**, for spe-

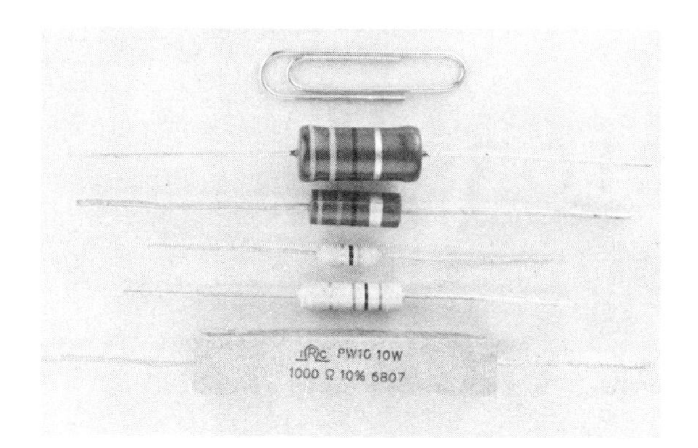

Figure 6-2 Resistors

cial thermostatic switches; **carbon** for the contact brushes in electrical motors and generators; and **tungsten** for the filaments of incandescent lamps.

Figure 6-3 Switch Contacts—Extreme Close-up

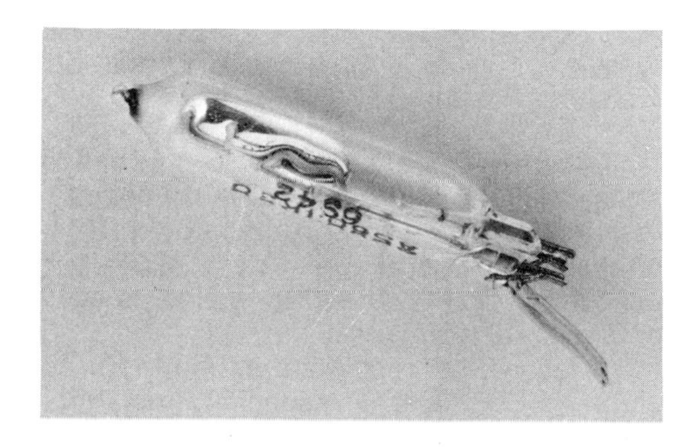

Figure 6-4 Mercury Switch for Thermostat

6-2 POOR CONDUCTORS

Poor conductors can be **dangerous** because they can conduct electricity to places where you do not expect to find it. Wet clothes, damp wood, moist earth, green plants, soiled rope, wet paper and many other materials belong in this group.

The **electrolytes** in primary and secondary cells also are classified as poor conductors; they are called **conducting solutions**.

6-3 INSULATORS

Substances that contain **no free electrons** under normal conditions cannot conduct electricity; they are called insulators, or insulating materials.

The best insulators are the **ceramic materials** such as **porcelain, steatite, titania** and others. These compounds can be poured or pressed into any desired shape and then fired to tremendous hardness in special furnaces. Ceramics are used wherever high-voltage and high-quality insulation is required.

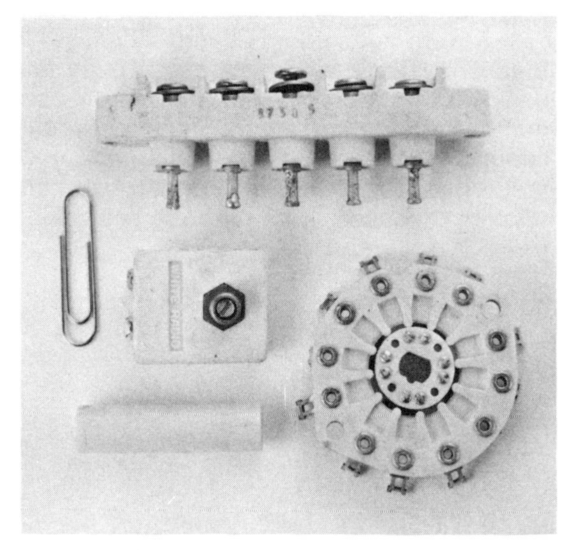

Figure 6-5 Ceramic Insulators

Mica is a mineral found in the form of thin, flexible sheets. It is high-temperature resistant and an excellent insulator. Mica is used extensively in electric heating devices.

Thermoplastic materials such as **nylon, teflon** and **vinyl** are excellent insulators. They are available in many brilliant colours. Most modern wires and cables are insulated with thermoplastics. The only disadvantage of these materials is their **low melting point**, which makes exposure to high temperatures an ever-present danger.

Asbestos, a grayish mineral consisting of long, thread-like fibres, is **fire-resistant** and an excellent insulator. It finds its chief use in electrical furnaces and in heat-resistant cables and appliance cords.

Many other materials also have good insulating characteristics. Among these are glass, dry air, mineral oil, dry paper, paraffin and rubber.

6-4 PRACTICAL CONDUCTORS

Most practical conductors are made of copper or aluminum in the shape of round, solid or stranded wires insulated with thermoplastic material in various colours. Only the wire sizes listed in the **American Wire Gauge**, or **AWG**, are available for general wiring purposes. The most common AWG sizes are shown in Figure 6-6.

ACTUAL CROSS-SECTION OF COMMON WIRE SIZES

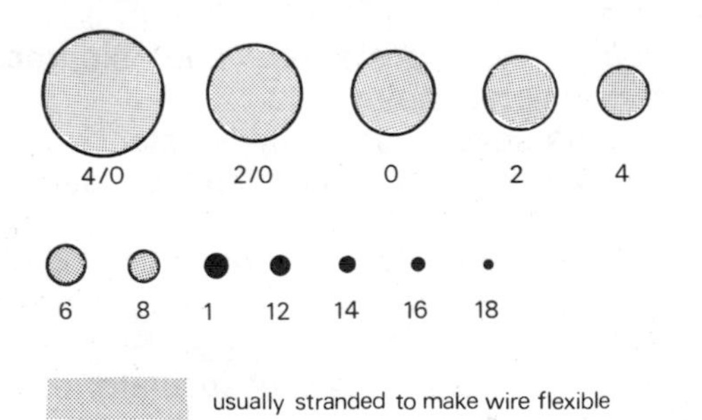

Figure 6-6 Actual Cross-sectional Area of Common Wire Sizes

To determine the **size** of a wire, the insulating sleeve is removed and the bare conductor is inserted into the fitting slot of a standard wire gauge, as shown in Figure 6-7.

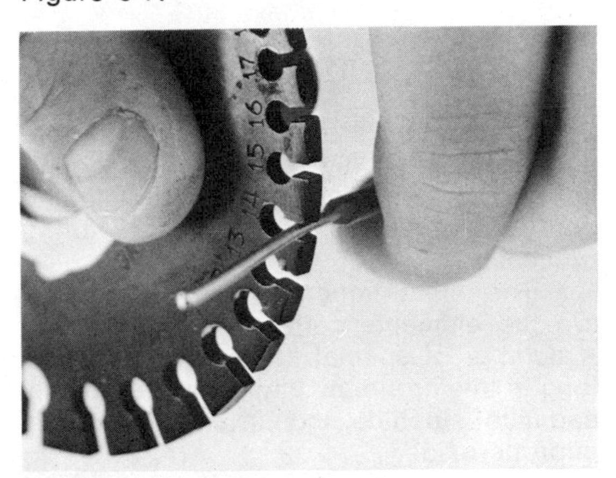

Figure 6-7 Using a Wire Gauge

The most common wire sizes used in the home are solid **AWG 14** for general-purpose wiring and stranded AWG 14 or 16 for lamp and appliance cords.

Both copper and aluminum wires are available in solid and in stranded form. Wires of size 6 and larger are always stranded to make them easier to work with. Single wires may be used only inside steel pipes called **conduits**. In house-wiring, the so-called **non-metallic sheath cable**, or **NMSC**, is preferred because it is both less expensive than conduit and easier to install.

Figure 6-8 Non-Metallic Sheath Cable

Coils for transformers, solenoids, motors and generators are wound with special **magnet wire**. This wire has a thin, tough and flexible, heat-resistant insulation that will not peel or crack. Common magnet wire sizes range from AWG 18 to AWG 30.

6-5 SEMICONDUCTORS

Semiconductors are substances that conduct electricity only under special conditions, or when "doped" with certain impurity materials during the manufacturing process.

The most important semiconductors are the elements germanium and silicon. A number of special compounds also belong in this group, mainly gallium arsenide, cadmium sulphide, indium antimonide and cuprous oxide.

A **pure** semiconductor is an **insulator** at low temperatures and if a low voltage is applied to it. If either temperature or voltage, or both, exceed a **critical value** the material will suddenly become a conductor. The extra heat energy or voltage tears electrons out of the semiconductor atoms and makes them available for conduction. When the temperature or voltage again drops below the critical point, the semiconductor returns into its insulator state.

Pure semiconductor materials are seldom used in practical devices.

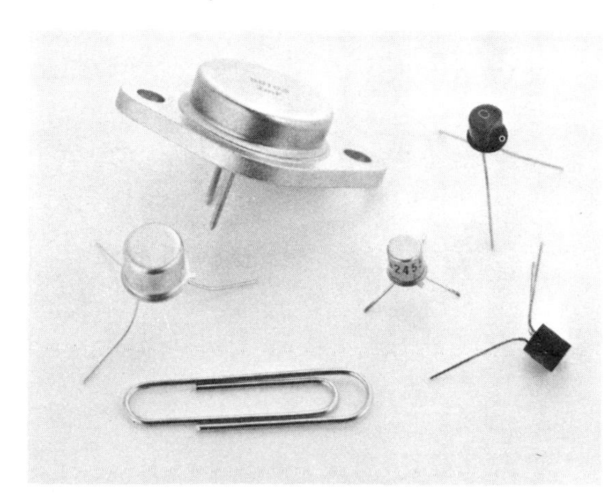

Figure 6-9 Semiconductor Devices

A **doped** semiconductor contains a carefully controlled amount of an impurity substance, or **dopant**. This impurity provides **charge carriers** in the otherwise non-conducting semiconductor. As a result, the doped semiconductor becomes a conductor. Depending on the impurity used, the semiconductor will be either a **P-type** or **N-type**, containing either positive or negative charge carriers.

Entirely new devices become possible when thin layers of P-type and N-type semiconductor material are formed into **P-N junctions**. In single or multiple junctions of this type, the flow of charge carriers can be **controlled** in many ways. Diodes, transistors, integrated circuits, solid-state lasers and light-emitting junctions are a few of the many semiconductor devices based on such P-N junctions.

6-6 ASSIGNMENTS: CONDUCTORS, SEMICONDUCTORS AND INSULATORS

6-6A Write Full Answers

1. List the four best conductors and state their most common uses.
2. Explain why gold is plated on the contact surfaces of connectors for computers and precision instruments.
3. Define the term specific resistance and state its value for silver, copper, gold and aluminum.
4. Why are constantan and nichrome called resistor materials?
5. Name several uses for resistance wire.
6. Why can poor conductors be dangerous to life?
7. List five poor conductors.
8. What do all insulator materials have in common?
9. Name three ceramic insulators and describe their general characteristics.
10. What is mica, and what are some of its applications?
11. State the names of three common

thermoplastic insulators, name their principal use, and describe their only serious drawback.

12. Why is asbestos the preferred insulating material in flexible line cords for irons and other heat-producing appliances?

13. List ten good insulator materials and their most common uses.

14. Name the most common wire types and sizes used in the home.

15. What is NMSC, and where is it used?

16. Describe the characteristics of magnet wire and state its main uses.

17. Why is a doped semiconductor a conductor?

18. How can a pure semiconductor be turned into a temporary conductor?

6-6B Indicate True or False

1. There is no clear dividing line between conductors and semiconductors.

2. Copper is the most common conductor material, followed by aluminum.

3. The specific resistance of a conductor is based on a wire one foot in length, regardless of diameter or shape.

4. Constantan and nichrome have a very high specific resistance.

5. Insulator materials do not conduct electricity because they have no free electrons.

6. Thermoplastic materials work well under all conditions.

7. Most practical conductors are made of aluminum, tungsten or copper.

8. Tungsten is used for the filaments of incandescent lamps.

9. There are many semiconductor elements.

10. Doped semiconductors contain either positive or negative charge carriers, depending on the type of impurity used during the production process.

6-6C Select Correct Answer

1. The best practical conductor is
   a) aluminum
   b) copper
   c) tungsten
   d) silver
2. The specific resistance of a conductor
   a) is of no significance to the technician
   b) is measured in ohms per foot
   c) is listed in the American Wire Gauge
   d) is measured in ohms per circular mil-foot
3. Poor conductors are dangerous because
   a) they conduct electricity to unexpected locations
   b) they do not conduct as well as copper or aluminum
   c) they conduct electricity with ease
   d) of their high specific resistance
4. The electrolytes in primary and secondary cells
   a) have excellent conductivity
   b) are classified as good conductors
   c) are conducting solutions
   d) can only exist in liquid form
5. Good insulators
   a) conduct electricity as well as poor conductors
   b) are found only among the ceramic materials
   c) all are heat resistant
   d) contain few or no free electrons
6. The preferred insulating material for electric furnaces is
   a) nylon
   b) asbestos
   c) vinyl
   d) teflon
7. The American Wire Gauge
   a) is an instrument for measuring conductors
   b) is a system of standard wire sizes
   c) is used only for solid conductors
   d) is not applicable to all practical conductors
8. The most common wire size used in the home is
   a) AWG 14
   b) AWG 18
   c) AWG 10
   d) AWG 12

9. A pure semiconductor
   a) conducts electricity at all times
   b) conducts only at low temperatures
   c) conducts when both temperature and voltage exceed a critical level
   d) behaves like an insulator at all times
10. P-type and N-type semiconductors
   a) have identical kinds of charge carriers
   b) are created by different impurities
   c) are identical to pure semiconductors in their electrical characteristics
   d) contain no charge carriers under any conditions

7

BASIC ELECTRIC CIRCUITS

Electric circuits can be as simple as the circuit in a toaster or in a lamp, or as complex as the central memory and control unit of a modern elevator with its hundreds of connections, relays and control devices.

Figure 7-1 Elevator Control Unit

Yet all the billions of circuits used in homes and in industry belong to only **five basic circuit types**: simple circuits, series circuits, parallel circuits, series-parallel circuits and complex circuits.

In the **simple electric circuit**, there is only one energy source, one control, and one load. As circuits become more complex, they generally still have only one power supply, but the number of loads and controls increases. Whether electrical components form a series, parallel or complex circuit depends on **how the loads are connected** to each other and to the supply.

7-1 THE SERIES CIRCUIT

A series circuit has **two or more loads** connected in single file, or in series, across a common power supply. In such a circuit, there is only a **single path** for the electrons, and if the circuit is broken at any point, no current will flow.

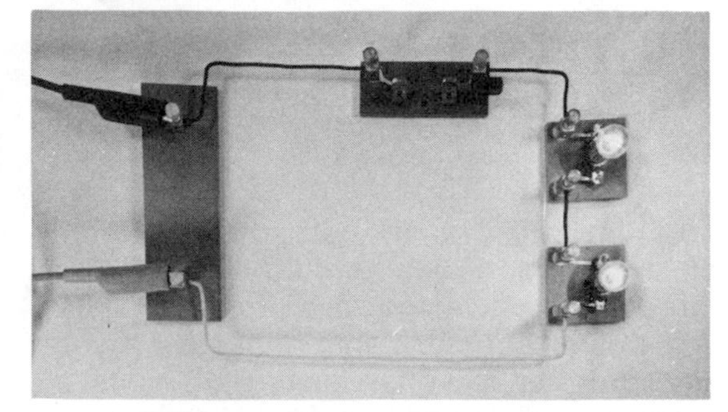

Figure 7-2 Basic Series Circuit

If both loads are identical, the free electrons will lose half of their total energy in each load, and the lamps will glow with equal intensity.

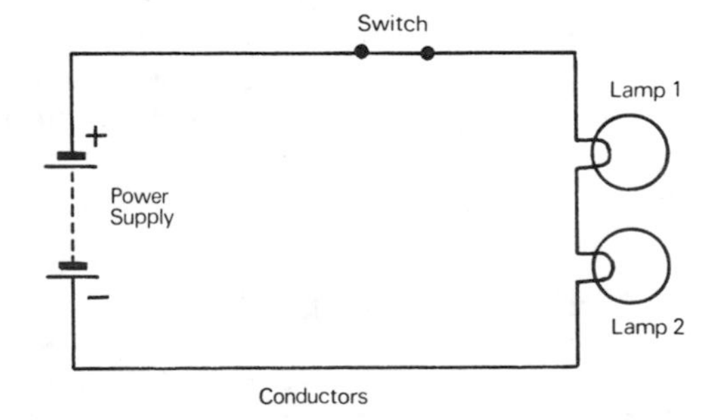

Figure 7-3 Basic Series Circuit—Schematic Diagram

If the loads are unequal, the electrons will lose more energy in the lamp which has the higher resistance, and that lamp will glow more brightly.

The **loss of voltage** (pressure) which the electrons experience in the loads is called a **voltage drop**. In a series circuit, the sum of the voltage drops around the circuit is always equal to the applied voltage. This statement is known as **Kirchhoff's Voltage Law.**

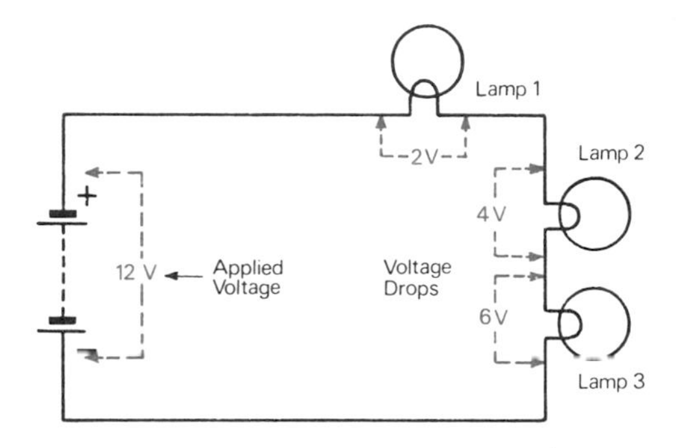

Figure 7-4 Typical Series Circuit

Figure 7-4 is the schematic diagram of a typical series circuit consisting of three different lamps, a switch, and a 12-volt power supply. Lamp 1 causes a voltage drop of 2 volts, leaving 10 volts of pressure for lamps 2 and 3. Lamp 2 decreases the force driving the electrons by another 4 volts, and lamp 3 uses up the remaining 6 volts. The free electrons then receive new energy from the supply for another round trip.

Series circuits are seldom used in the lighting, heating and appliance circuits of a home. But in complex equipment such as TV receivers, computers and automated machinery, series circuits are of vital importance.

7-2 THE PARALLEL CIRCUIT

A parallel circuit has **two or more loads**, each connected directly to the power supply. Thus, each load forms an individual circuit with the supply, and all loads share the conductors leading to and from the energy source.

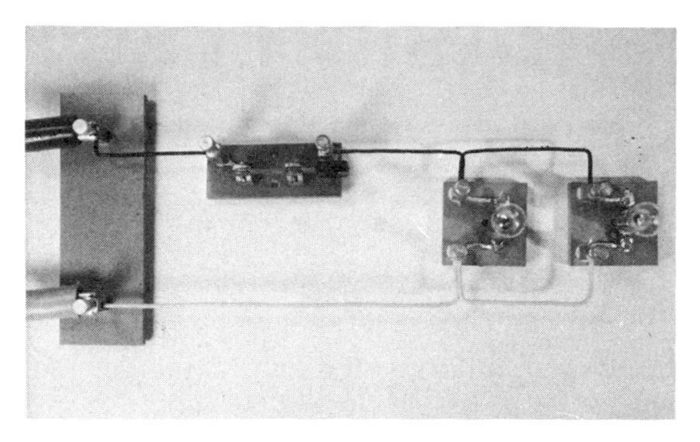

Figure 7-5 Basic Parallel Circuit

Each load forms a **branch circuit**, sharing the supply with the other loads. These branch circuits operate **independently**. Each load receives the entire voltage of the supply and all the energy of those electrons flowing through it, and removal of a load will not affect the others.

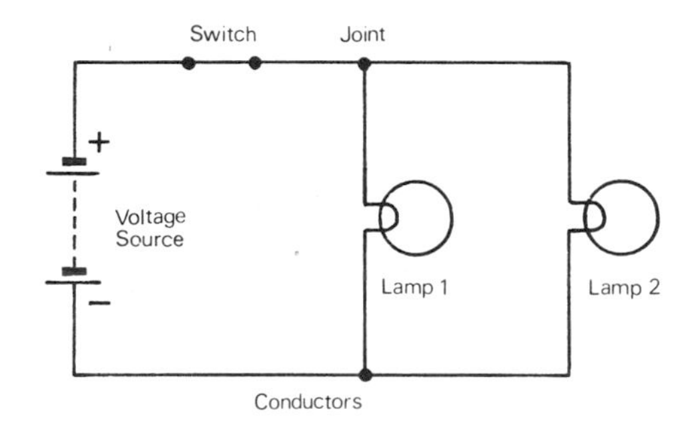

Figure 7-6 Basic Parallel Circuit—Schematic Diagram

Since each load is connected directly across the supply, each receives the **full supply voltage**. The total circuit current flowing through the supply is equal to the sum of the individual load currents. This statements is known as **Kirchhoff's Current Law.**

In the parallel circuit shown in Figure 7-7, three different-size lamps are connected across a 12-volt supply. Lamp 1 draws 0.5 amperes from the supply; lamp

2, 1.5 amperes; and lamp 3, 2 amperes. Each lamp receives the full 12 volts, and the total circuit current is 4 amperes.

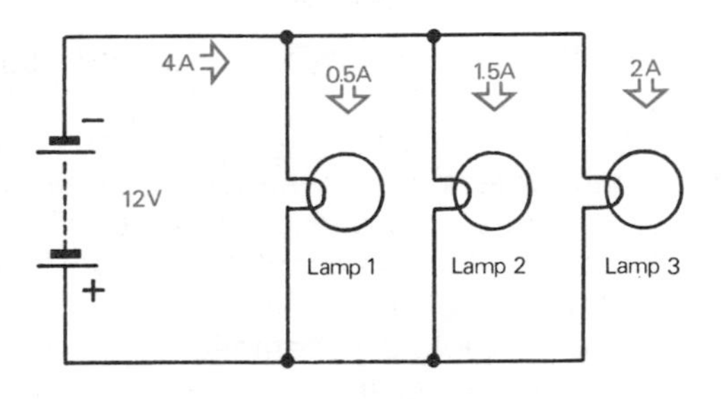

Figure 7-7 Typical Parallel Circuit

The parallel circuit is the most commonly used circuit because all loads have the same voltage rating and can be operated independently.

The electrical system of a home can be seen as a single parallel circuit with many branch circuits and a common 115-volt a.c. supply. That is why all standard home appliances are designed to work with 115 volts a.c.

7-3 THE SERIES-PARALLEL CIRCUIT

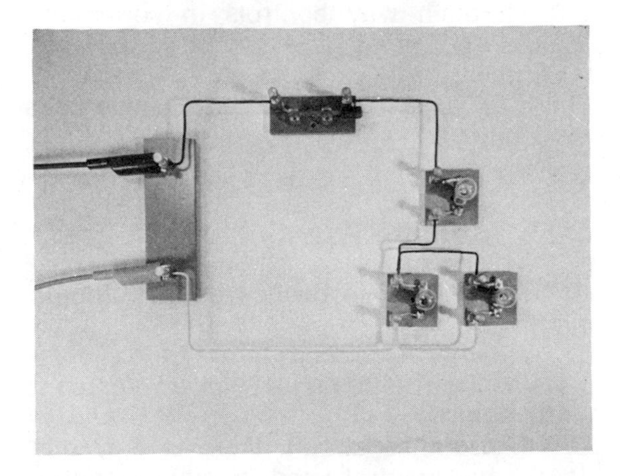

Figure 7-8 Basic Series-Parallel Circuit

This circuit is a **combination** of **series** and **parallel** circuit elements connected to a common power supply. Figures 7-8 and 7-9 show the simplest series-parallel circuit possible.

If lamps 2 and 3 are equal, the electron current will divide into two equal branch currents. The two parallel lamps are themselves in series with lamp 1, which receives the total current provided by the power supply.

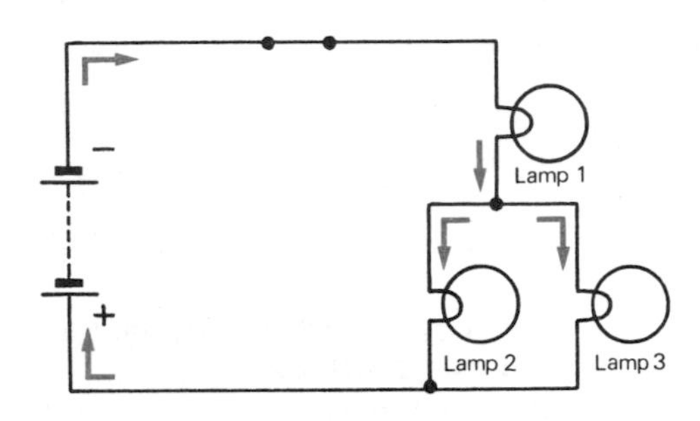

Figure 7-9 Basic Series-Parallel Circuit — Schematic Diagram

If lamp 1 is removed or burns out, the entire circuit is broken, and no current will flow. If either lamp 2 or lamp 3 is removed, the remaining lamps will receive more energy and glow more brightly.

The voltage across lamps 2 and 3 is common to both lamps (the same), and the voltage across lamp 1 and the parallel lamps 2 and 3 is equal to the supply voltage.

Series-parallel circuits are not used in the home, but are quite common in industrial electrical and electronic equipment.

7-4 CONTROLS IN SERIES AND IN PARALLEL

In many electric circuits, more than one switch is used to control the current flow. Controls can be connected in series or in parallel, depending on the kind of operation planned for the circuit.

Controls in series are used when a common load must operate only when two or more switches are closed simultaneously. Power-operated shears are a typical example. The shears will work only if both switches are pushed at the same time, thus keeping the operator's hands out of harm's way.

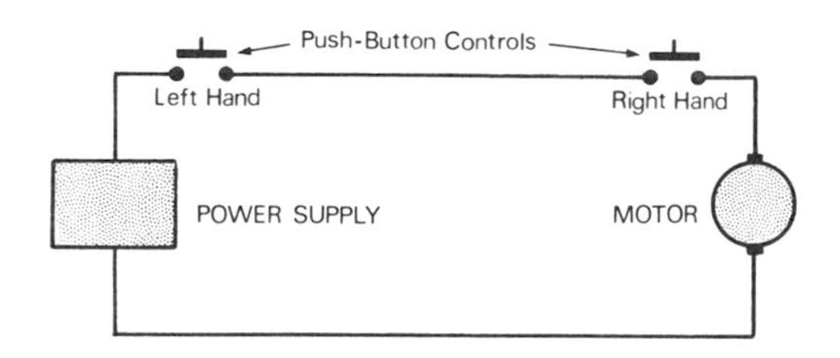

Figure 7-10 Series Controls on Power Shears

In electronics, switches connected in series are called **AND gates**, since both switch A **and** switch B must be operated simultaneously to close the circuit and permit the current to flow.

Controls in parallel are used when a common load is to be turned on or off from two or more locations. The simple bell circuit in Figure 7-11 is a typical example.

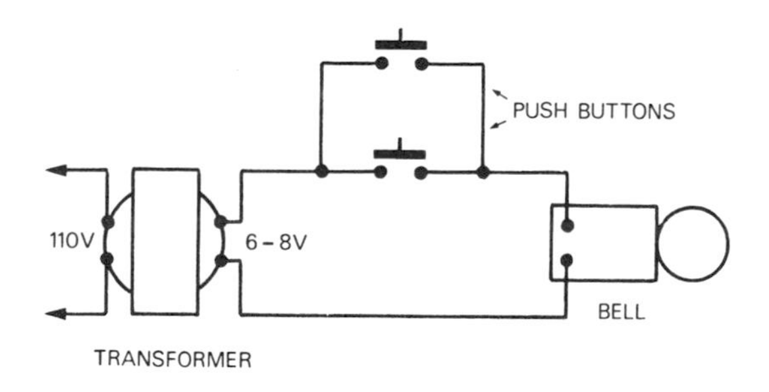

Figure 7-11 Parallel Controls in Simple Bell Circuit

In this circuit, the bell can be rung by pushing either one, or the other, or both push-buttons.

In electronics, controls in parallel are called **OR gates**, because either switch A **or** switch B can control the current flow in the circuit.

7-5 ASSIGNMENTS: BASIC ELECTRIC CIRCUITS

7-5A Write Full Answers

1. List the five basic types of electric circuits.
2. Describe the simple electric circuit.
3. What determines whether a circuit is either a series or a parallel circuit?
4. Describe the physical and electrical characteristics of the basic series circuit.
5. Draw a labelled schematic diagram of the basic series circuit.
6. Explain what happens to the free electrons as they flow around a series circuit.
7. Describe the physical and electrical characteristics of the basic parallel circuit.
8. Draw a labelled schematic diagram of the basic parallel circuit.
9. Explain how the applied voltage and the total current are distributed in a parallel circuit.
10. What happens in the basic series-parallel circuit when any one of the loads is removed?
11. Describe the voltage distribution in the series-parallel circuit.
12. State at least two good reasons why the series-parallel circuit is not used in the home.
13. Explain why controls in series are called AND gates, and controls in parallel, OR gates.
14. Try to discover several effective uses for controls in series.

7-5B Indicate True or False

1. There are five basic types of electric circuits.
2. Whether a circuit is a series or a parallel type depends on the arrangement of its controls.
3. In a simple circuit, there is only one load, one control, and a single energy source.
4. If one load is removed from a series circuit, the other loads still work.

5. In each load in a series circuit, the electrons lose a part of their total kinetic energy.
6. In a series circuit, the load with the greatest resistance causes the smallest voltage drop.
7. In a parallel circuit, each load has a separate power supply.
8. The total current flowing in a parallel circuit is equal to the sum of its branch currents.
9. In a parallel circuit, all loads receive different voltage levels.
10. In a series-parallel circuit, each load depends on all other loads for its voltage and current level.
11. The series-parallel circuit is widely used in homes and industry.
12. In a series-parallel circuit, all loads receive the same amount of electron current.
13. Controls in series are called AND gates because switch A and switch B must be operated simultaneously.
14. Controls in parallel are called OR gates because either switch A or switch B will operate the load.

7-5C **Select Correct Answer**

1. As circuits become more complex,
   a) not all electron paths return to the power supply
   b) the number of supplies and controls increases
   c) they still have only one path for electrons
   d) the number of loads and controls increases
2. A series circuit
   a) has two or more loads and several power supplies
   b) has two or more loads connected in single file
   c) has more than one path for electrons
   d) works only with identical loads
3. In a series circuit,
   a) each load receives a different current flow
   b) the energy received by a load is independent of the load's resistance
   c) the voltage lost in a load is called a voltage rise
   d) the sum of the voltage drop is equal to the applied voltage
4. In a parallel circuit, the individual current paths
   a) depend on one another
   b) do not return to the common supply
   c) are called branch circuits
   d) carry the total circuit current
5. The total current in a parallel circuit
   a) is the sum of the branch currents
   b) is independent of the branch currents
   c) flows through all loads
   d) is independent of the loads
6. All loads in a parallel circuit
   a) receive different voltage levels
   b) are connected across the full supply voltage
   c) are connected in single file
   d) share each other's electron current
7. In a series-parallel circuit,
   a) each load depends on all other loads for its voltage and current level
   b) all loads receive the same amount of electron current
   c) the total current flows through all loads
   d) the current in each load is independent of the other load currents
8. The series-parallel circuit
   a) is commonly used in home appliance circuits
   b) has a single path for free electrons
   c) has more than one power supply
   d) combines the physical and electrical characteristics of the series and the parallel circuit
9. Controls in series
   a) are called OR gates
   b) must be operated simultaneously to close the circuit

c) provide two or more paths for the electron current
d) are not used in industry
10. Controls in parallel
a) are called OR gates
b) provide a single path for free electrons
c) must be operated simultaneously
d) serve no practical purpose

8
OHM'S LAW

In 1827, a professor of physics, George S. Ohm, published the results of his research into the behaviour of dynamic electricity. Using instruments invented and built by himself, Professor Ohm had discovered a **mathematical relationship** between the **voltage, current** and **resistance** in an electric circuit. Ohm summarized his work in a simple equation, now known as **Ohm's Law**:

$$I = \frac{E}{R}$$

This equation has become the most essential tool for working with electric and electronic circuits. It is valid for both a.c. and d.c. circuits.

8-1 ELECTRON CURRENT VS. APPLIED VOLTAGE

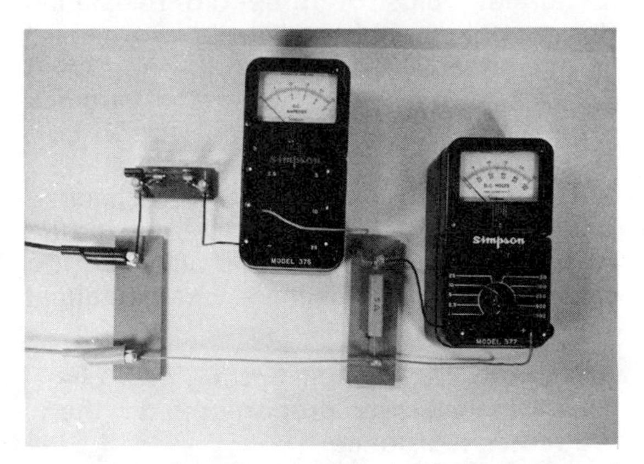

Figure 8-1 Experimental Circuit: Electron Current vs. Applied Voltage

Figures 8-1 and 8-2 show the circuit used to determine the validity of the first part of Ohm's Law, the relationship between electron current and applied voltage.

A variable voltage d.c. supply is connected to a wire-wound resistor with a constant resistance of 5 ohms. Voltages of 1.5 V, 3.0 V, 4.5 V, 6 V and 7.5 V are then applied, and the resulting electron current is measured each time.

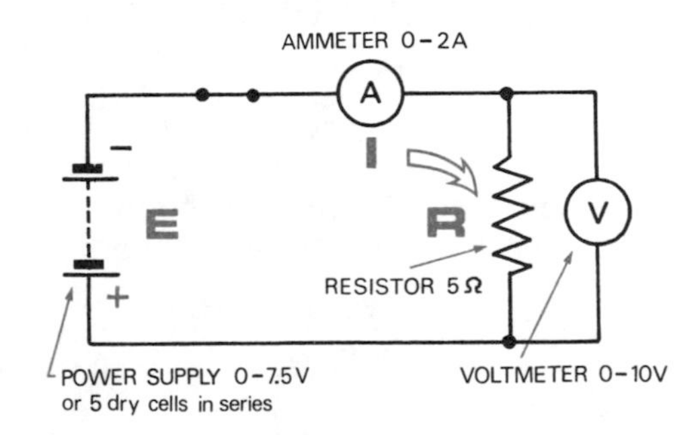

Figure 8-2 Experimental Circuit: Electron Current vs. Applied Voltage—Schematic Diagram

The results of the experiment are listed in Table 8-1.

APPLIED VOLTAGE	RESULTING CURRENT	CIRCUIT RESISTANCE
1.5V	0.3A	
3.0V	0.6A	
4.5V	0.9A	constant (5 ohms)
6.0V	1.2A	
7.5V	1.5A	

Table 8-1 Electron Current vs. Applied Voltage

Here is an analysis of the results:

1. When the applied voltage is doubled (from 1.5 V to 3 V), the electron current also doubles.
2. When the applied voltage is tripled or quadrupled, the electron current also increases to three or four times its original value.
3. Any decrease in the applied voltage (reading the table from the bottom to the top) would result in a proportional decrease of the electron current.

The experiment proves the existence of a mathematical relationship between electron current and applied voltage. Its results can be summarized in a simple statement:

The electron current in a circuit is directly proportional to the applied voltage.

This is the first part of Ohm's Law.

8-2 ELECTRON CURRENT VS. RESISTANCE

Figures 8-3 and 8-4 show the circuit used in an experiment to determine how the electron current flowing in a circuit is influenced by the circuit's resistance.

Figure 8-3 Experimental Circuit: Electron Current vs. Resistance

A constant d.c. voltage of 1.5 volts is applied to a circuit with a 1-ohm load, and the electron current is measured. The load resistance is then changed to 2, 3, 4 and 5 ohms, respectively, and the new current level is measured each time.

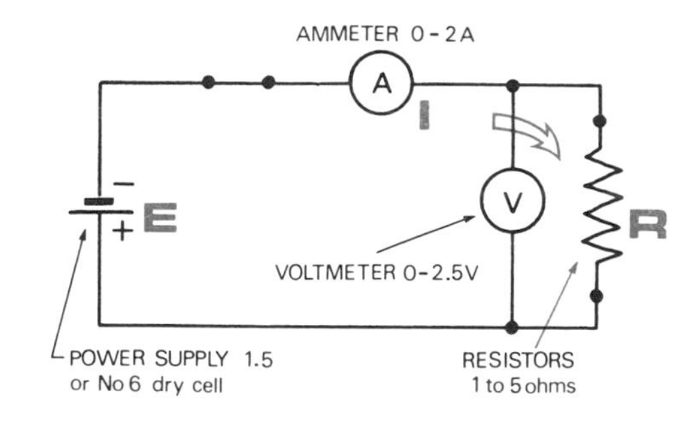

Figure 8-4 Experimental Circuit: Electron Current vs. Resistance—Schematic Diagram

The results of the experiment are tabulated in Table 8-2.

APPLIED VOLTAGE	RESULTING CURRENT	CIRCUIT RESISTANCE
	1.50A	1 ohm
	0.75A	2 ohms
constant (1.5 volts)	0.50A	3 "
	0.37A	4 "
	0.50A	5 "

Table 8-2 Electron Current vs. Resistance

Each time the circuit's resistance is doubled, the electron current drops to half its former value. A three-fold resistance (from 1 ohm to 3 ohms) results in one-third of the previous current level (from 1.5 A down to 0.5 A). Multiplying the circuit's resistance by five drops the electron current to one-fifth of its original value.

Generally, any decrease in resistance leads to a proportional decrease in the electron current. These results can be summarized in a simple mathematical statement:

The electron current flowing in a circuit is inversely proportional to the circuit's resistance.

This is the second part of Ohm's Law.

8-3 OHM'S LAW

In the two preceding sub-chapters, we actually repeated George Ohm's original experiments, although on a much smaller scale and with much better equipment. We can now combine the results of both experiments into a single conclusion, a verbal form of **Ohm's Law**:

> **The electron current in a circuit is directly proportional to the applied voltage, and inversely proportional to to the circuit resistance.**

This verbal form of the law can be translated into a mathematical equation, using the symbols I, E and R for current, voltage and resistance:

$$I = \frac{E}{R}$$

where I is current in amperes
E is voltage in volts
and R is resistance in ohms

Figure 8-5 Ohm's Law

This **basic form** of Ohm's Law can be used to calculate the amount of electron current flowing in a circuit if the applied voltage and the circuit resistance are known. Here is a sample calculation:

The heating element of a toaster has a resistance of 8 ohms. How much current will this toaster draw from a 116-volt supply?

1. $E = 116$ volts; $R = 8$ ohms; $I = ?$
2. $I = \frac{E}{R}$
3. $I = \frac{116}{8}$
4. $8\overline{)\ 116}$ 14.5
5. $\mathbf{I = 14.5}$ **amperes**

The basic form of Ohm's Law can be changed by simple mathematical operations to obtain two **alternate forms**:

$I = \frac{E}{R}$ (basic equation)

$I \times R = \frac{E \times R}{R}$ (multiply both sides by R)

$I \times R = \frac{E \times \cancel{R}}{\cancel{R}}$ (eliminate common factors)

$\mathbf{E = I \times R}$ (first derived form of Ohm's Law)

Another sample calculation shows how this form is used in practical situations:

How much voltage will be lost across a 4.2-ohm resistor if a current of 3 amperes flows through it?

1. $I = 3$ amperes; $R = 4.2$ ohms; $E = ?$
2. $E = I \times R$
3. $E = 3 \times 4.2$
4. $\mathbf{E = 12.6}$ **volts**

Using the first derived form of Ohm's Law, a **second** alternate equation can be developed:

$E = I \times R$ (first derived form)

$\frac{E}{I} = \frac{I \times R}{I}$ (divide both sides by I)

$\frac{E}{I} = \frac{\cancel{I} \times R}{\cancel{I}}$ (eliminate common factors)

$\mathbf{R = \frac{E}{I}}$ (second derived form of Ohm's Law)

Once again, here is a sample calculation to show you how this equation is put to work:

Find the resistance that will limit the electron current in a circuit to 4.6 amperes if a voltage of 115 volts a.c. is applied.

1. $E = 115$ volts a.c.; $I = 4.6$ amperes; $R = ?$
2. $R = \frac{E}{I}$
3. $R = \frac{115}{4.6}$
4. $4.6\overline{)\ 115}$ 25
5. $\mathbf{R = 25}$ **ohms**

Simple memory devices can be used to help you remember all three equations of Ohm's Law:

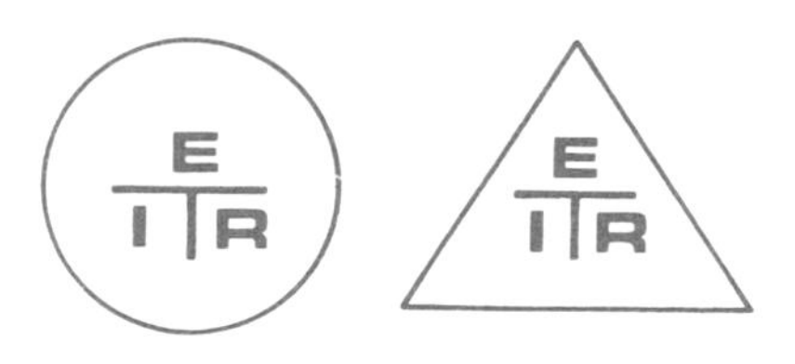

Figure 8-6 Memory Devices for Ohm's Law

If you cover the quantity you wish to calculate (voltage, current, or resistance), the remaining parts of the memory circle or triangle will show you which form of Ohm's Law you must use:

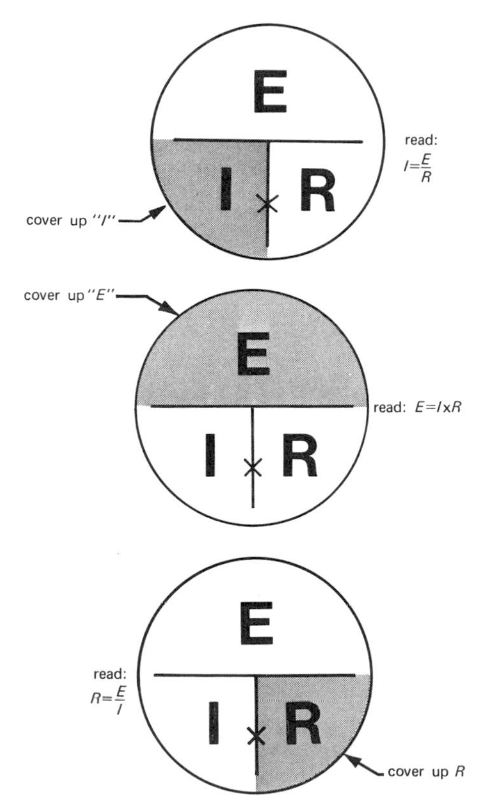

Figure 8-7 Using the Memory Circle for Ohm's Law

These memory devices are not Ohm's Law; they merely show a convenient method for remembering the three forms of the law.

8-4 ASSIGNMENTS: OHM'S LAW

8-4A Calculating Electron Current

1. Calculate the electron current flowing through a 4-ohm resistor if a voltage of 6 volts is applied across its terminals.
2. An incandescent lamp with a hot resistance of 90 ohms is connected across the 115-volt a.c. line. How much current will flow?
3. A transistor radio with an internal resistance of 450 ohms is operated with a 9-volt battery. How much current must the battery supply?
4. The heating element of a soldering iron has an internal resistance of 200 ohms. How much current will the device draw from the 115-volt line?
5. The cranking motor of an automobile has a resistance of 0.12 ohms. How much current must a 12.6-volt storage battery supply to start the car?
6. Calculate the electron current flowing through a light bulb with a resistance of 3 ohms, if the bulb is connected to a 6.3-volt battery.
7. A photovoltaic cell (solar cell) generates a voltage of 0.45 volts in full sunlight. How much current will this cell drive through a 9-ohm load?
8. A citizens band radio with an internal resistance of 250 ohms is connected to a 6-volt supply. Calculate the electron current necessary to operate the set.
9. A piezoelectric crystal generates 1.15 volts when twisted by a phonograph needle. How much current will the crystal drive through the 200,000-input resistance of an amplifier?
10. Calculate the electron current for each of the five problems listed below. Use correct mathematical procedure in each case:

a) E = 0.2 volts; R = 16 ohms

b) E = 72 volts; R = 24 ohms
c) E = 117 volts; R = 23.4 ohms
d) E = 12.6 volts; R = 3.15 ohms
e) E = 6 volts; R = 18 ohms

8-4B **Calculating Applied Voltage**

1. An electron current of 2 amperes flows through a 3.6-ohm wire-wound resistor. Calculate the applied voltage.
2. A light bulb with a hot resistance of 90 ohms needs a 1.3-ampere current for full brilliancy. Calculate the voltage required.
3. A transistor radio with an internal resistance of 200 ohms requires a current of 0.03 amperes for its operation. What battery voltage is needed?
4. How much voltage must be applied to the 50-ohm heating element of a soldering iron that requires an electron current of 2.3 amperes to reach proper temperature?
5. Calculate the battery voltage needed to drive a current of 200 amperes through the cranking motor of a car. The motor's resistance is 0.03 ohms.
6. A small relay requires a drive current of 0.025 amperes to close its switch contacts. Its coil has a resistance of 240 ohms. How much voltage is needed?
7. A solar cell provides an electron current of 0.15 amperes through a 30-ohm load. Calculate the output voltage of that cell.
8. How many carbon-zinc dry cells must be connected in series to produce a 0.5-ampere electron current in a 24-ohm load?
9. A piezoelectric crystal produces an a.c. current of one-millionth of an ampere through the eleven-million ohm resistance of a sensitive voltmeter. What voltage will the meter indicate?
10. Calculate the applied voltage for each of the problems listed. Use correct mathematical procedure in each case:

a) I = 0.005 amperes; R = 16,000 ohms
b) I = 17.2 amperes; R = 6.2 ohms
c) I = 20 amperes; R = 16.5 ohms
d) I = 0.033 amperes; R = 0.05 ohms
e) I = 8.2 amperes; R = 12 ohms

8-4C **Calculating Resistance**

1. Calculate the resistance of a circuit in which an applied voltage of 117 volts will produce an electron current of 2.5 amperes.
2. A light bulb is connected to the 115-volt a.c. line. Find its hot resistance if the electron current flowing through its filament is 0.46 amperes.
3. A small transistor radio requires 9 volts and 0.02 amperes to operate properly. Calculate its internal resistance.
4. Calculate the resistance of the heating element of a soldering iron which requires 110 volts a.c. and 1.25 amperes to reach soldering temperature.
5. The headlight of an automobile needs a current of 3 amperes and a supply voltage of 12.6 volts. Calculate its hot resistance.
6. Calculate the resistance of the combined heating elements of a toaster which draws 11 amperes from the 117-volts a.c. line.
7. A desk lamp with a current rating of 0.5 amperes at 110 volts is to be replaced by a wire-wound resistor. Calculate the value of that resistor.
8. Calculate the internal resistance of a coil which draws an electron current of 0.05 amperes from a fresh mercury cell.
9. A 12-volt battery causes an electron current of 1.25 amperes to flow through a car radio. Calculate the radio's internal resistance.
10. Calculate the amount of resistance required for the problems listed below. Use correct mathematical procedure in each case:

a) E = 117 volts; I = 0.25 amperes
b) E = 1.5 volts; I = 0.5 amperes
c) E = 60 volts; I = 2.5 amperes
d) E = 112 volts; I = 8 amperes
e) E = 0.45 volts; I = 0.2 amperes

8-4D **General Problems**

1. An electron current of 1.5 amperes flows through a 5-ohm resistor. Calculate the applied voltage.

2. An emf of 110 volts is applied to a 25-ohm resistor. Find the circuit current.

3. A Number 6 dry cell is momentarily short-circuited, and a 5-ampere current flows through the 0.05-ohm internal resistance of the cell. To what value does its voltage collapse?

4. Heat energy forces an electron flow of 0.15 amperes across the junction of a thermocouple into a 12-ohm load. Calculate the terminal voltage of the couple.

5. A moving magnet induces an electron current of 0.0005 amperes in a coil. Calculate the resistance of the coil if the induced voltage is 0.002 volts.

6. Calculate the missing quantities in the problems listed below. Be careful in the development of each answer, and show every step of your calculations:

	E	R	I
a)	E = 27V;	R = 81 ohms;	I = ?
b)	E = ?	R = 150 ohms;	I = 0.75A
c)	E = 117V;	R = ?	I = 9.2A
d)	E = ?	R = 405 ohms;	I = 0.25A
e)	E = 12.6V;	R = 6 ohms;	I = ?
f)	E = 1.5V;	R = 5 ohms;	I = ?
g)	E = ?	R = 55 ohms;	I = 5A
h)	E = 110V;	R = 25 ohms;	I = ?

9
ELECTROMAGNETISM

Early experimenters believed in the existence of some sort of relationship between electricity and magnetism. During the seventeenth and eighteenth centuries, individual researchers attempted to discover this link. They did not succeed.

In 1819, Hans Christian Oersted, a professor at the University of Copenhagen, discovered electromagnetism during a routine classroom experiment.

9-1 OERSTED'S EXPERIMENT

Oersted was demonstrating dynamic electricity to a group of students when he noticed that the needle of a nearby magnetic compass moved each time he closed the circuit. Oersted immediately realized that he had made a discovery of major importance. He refined the experiment and began a systematic investigation of this new form of magnetism. Today his work is considered one of the major advances in the development of electrical technology.

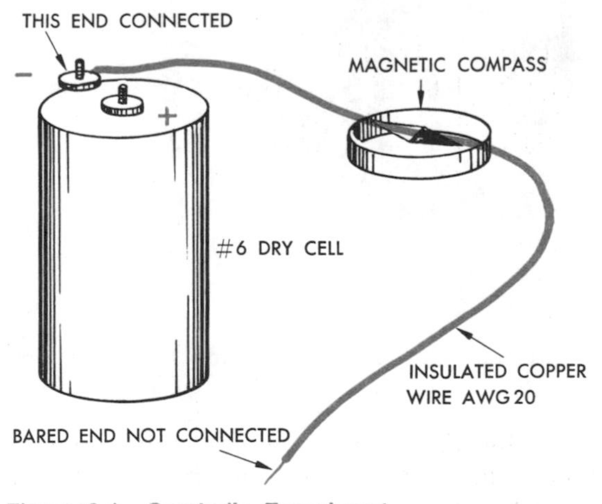

Figure 9-1 Oersted's Experiment

A Number 6 dry cell, about three feet (or a metre) of AWG 20 wire, and a magnetic compass are all we need to repeat **Oersted's original experiment.**

Figure 9-2 Oersted's Experiment — Close-up of Wire and Compass Needle

Put the magnetic compass on your desk, one foot or about 30 centimetres from the dry cell. Strip the insulation off both ends of the wire and connect one end to the negative terminal of the cell. Run the insulated wire across the top of the compass, in line with the needle (which will be in its normal, northseeking alignment). Now, while holding the wire in place with one hand, touch its free end

briefly to the positive cell terminal and observe the reaction of the compass needle. Repeat this several times.

Each time the circuit is closed and an electron current flows through the conductor, the compass needle is strongly **deflected** by a magnetic force. The non-magnetic copper wire does not affect the compass needle. Therefore, it must be the **electron current** itself which **generates the magnetic force**.

This experiment confirms the principal rule of electromagnetism: **free electrons moving through a conductor generate an invisible magnetic force.** This electromagnetic force disappears when the electron current ceases to flow.

9-2 CHARACTERISTICS OF ELECTROMAGNETIC FORCE

If the copper conductor used in Oersted's experiment is pushed through a piece of cardboard with iron filings sprinkled on it, and if the experiment is repeated, an iron-filings pattern similar to Figure 9-3 will be generated.

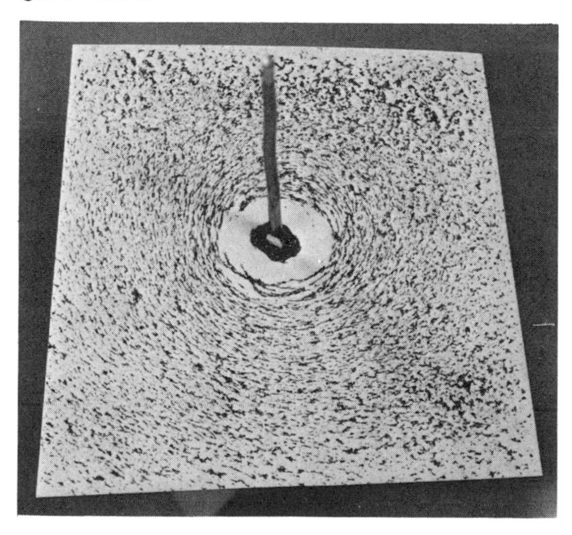

Figure 9-3 Iron-Filings Pattern Around a Current-Carrying Conductor

The alignment of the filings shows that the magnetic force of the electron current acts along **circular flux lines**, at right angles to the conductor. Each successive line completely encloses the previous one, and all have a **common centre**, the current-carrying conductor.

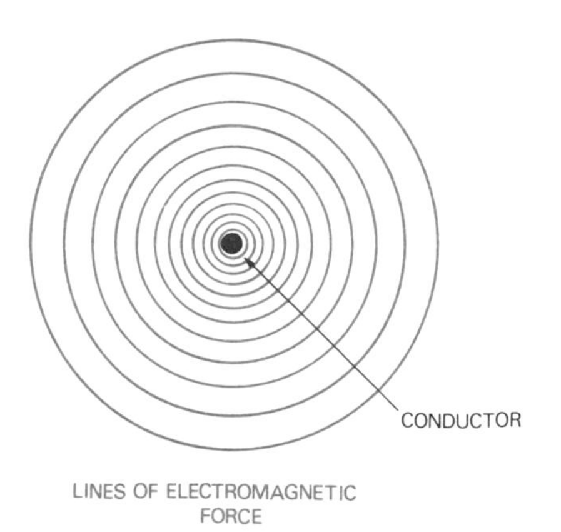

Figure 9-4 Lines of Force Around a Current-Carrying Conductor

If the experiment is repeated with several cardboard squares arranged as shown in Figure 9-5, each layer of iron filings will be aligned into a circular pattern. This proves that the electromagnetic force **acts along the entire conductor.**

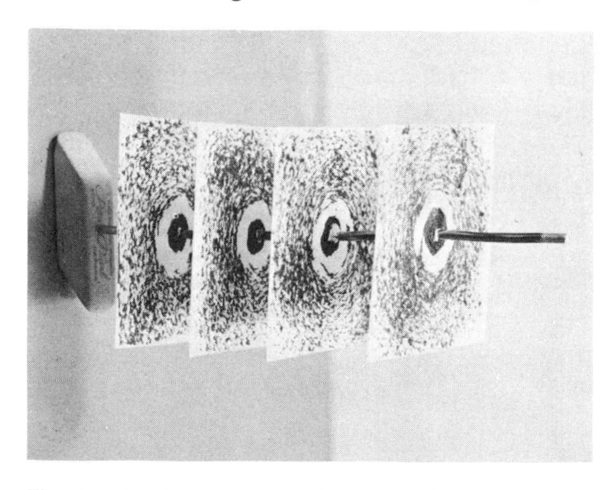

Figure 9-5 Electromagnetic Force Acts Along the Entire Conductor

The **direction** in which this electromagnetic force acts can be shown in several ways. If small magnetic compasses are set up around the current-carrying conductor, all their needles will be **aligned in the same direction**.

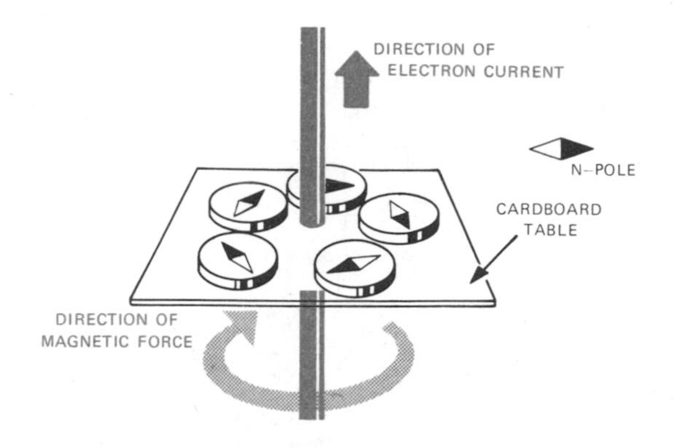

Figure 9-6 Magnetic Force Around a Current-Carrying Conductor—Clockwise Direction

Reversing the direction of electron flow through the conductor will also reverse the alignment of the compass needles. This proves that the direction of the electromagnetic force **depends** on the direction of the electron flow.

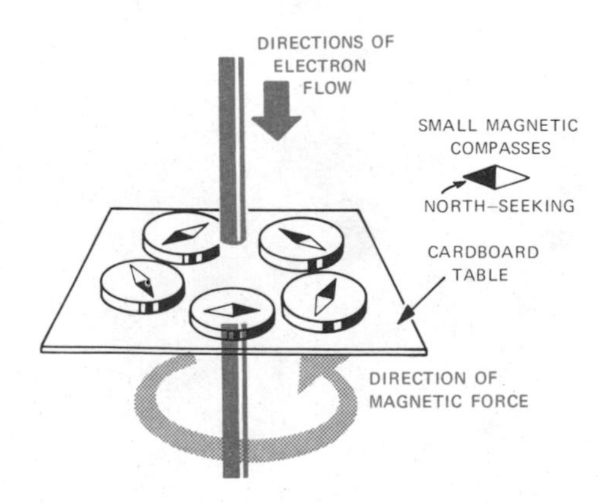

Figure 9-7 Magnetic Force Around a Current-Carrying Conductor—Counterclockwise Direction

To understand the relationship between the direction of electron flow and the direction of the resulting electromagnetic force **visualize** the cross-section of the conductor, with electrons flowing either away from you (tail of arrow), or toward you (tip of arrow), as shown in Figures 9-8 and 9-9. The lines of electromagnetic force act in the direction in which they **push a free N-pole**.

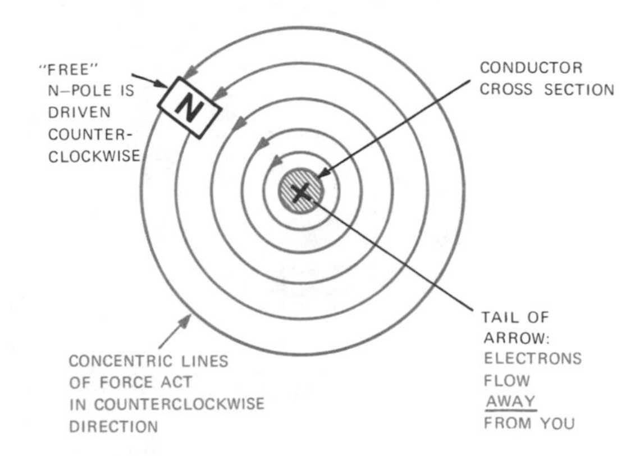

Figure 9-8 Direction of Electromagnetic Force—Electrons Flowing Away from You

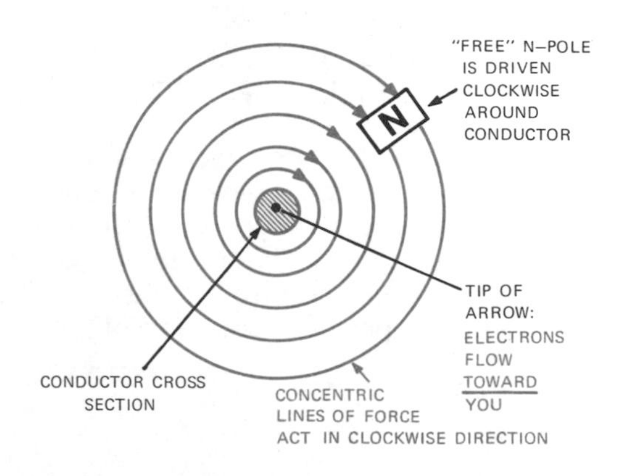

Figure 9-9 Direction of Electromagnetic Force—Electrons Flowing Toward You

The technician uses the **left-hand rule for single conductors** to determine the direction of an electromagnetic force:

> If the thumb of the **left hand** points along the conductor in the direction of the electron flow, the curled fingers point in the direction of the electromagnetic force.

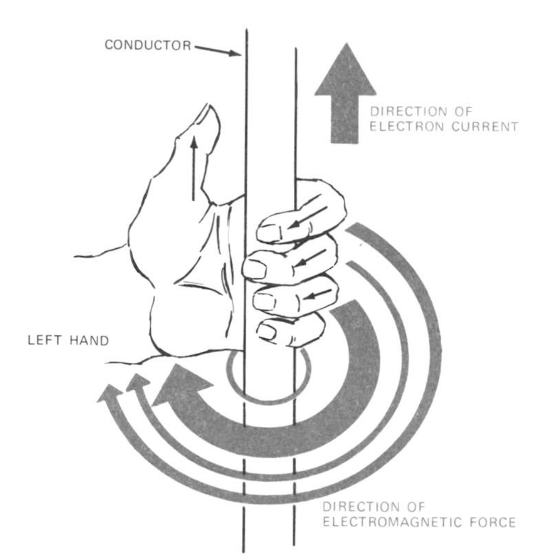

Figure 9-10 The Left-Hand Rule for Current-Carrying Conductors

9-3 ELECTROMAGNETIC FORCES BETWEEN PARALLEL CONDUCTORS

When two conductors, each carrying an electron current, come close to one another, their electromagnetic forces **interact**. Depending on the direction of the individual electron currents, the conductors either **attract or repel** each other.

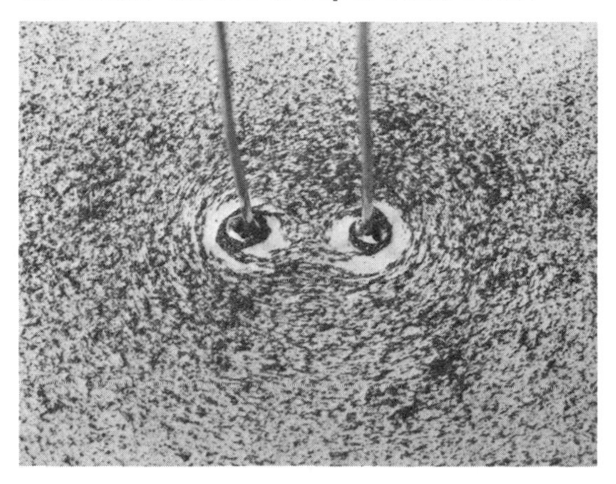

Figure 9-11 Iron-Filings Pattern Around Parallel Conductors, Currents Flowing in Same Direction

Mutual attraction occurs when both electron currents flow in the **same direction** through the conductors. The iron-filings pattern in Figure 9-11 shows that the flux lines around the conductors attract one another and **merge** into lines enclosing both conductors.

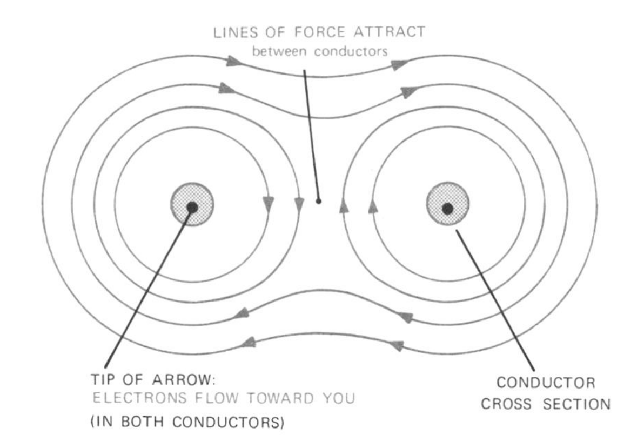

Figure 9-12 Lines of Electromagnetic Force Around Parallel Conductors, Currents Flowing in Same Direction

The flux lines act in a manner similar to the way rubber bands would pull the conductors together.

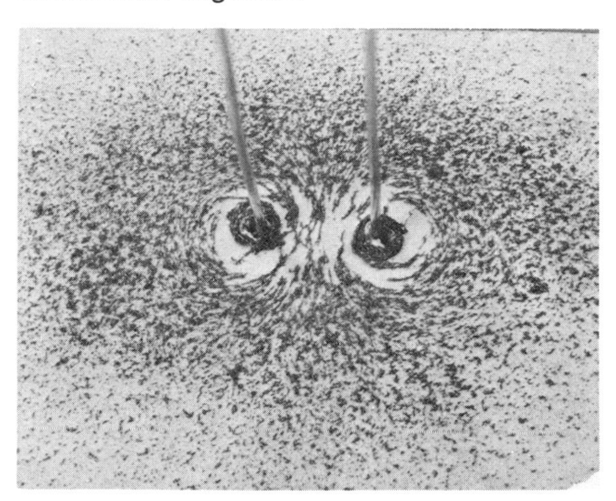

Figure 9-13 Iron-Filings Pattern Around Parallel Conductors, Currents Flowing in Opposite Direction

Mutual repulsion results when the electron currents flow in **opposite directions** through the parallel conductors. The iron-filings pattern in Figure 9-13 shows that each set of flux lines **repels** the other set, thus pushing the conductors apart.

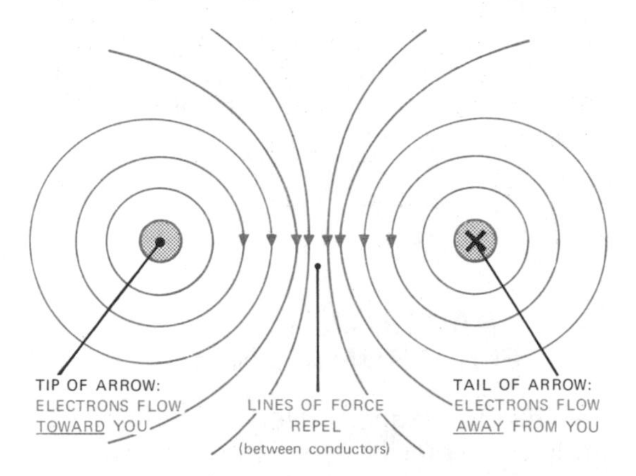

Figure 9-14 Lines of Electromagnetic Force Around Parallel Conductors, Currents Flowing in Opposite Direction

9-4 THE ELECTROMAGNET

If a current-carrying conductor is wound into the shape of a **coil**, the flux lines of the individual turns **merge** to form an **electromagnetic field** similar to that of a bar magnet.

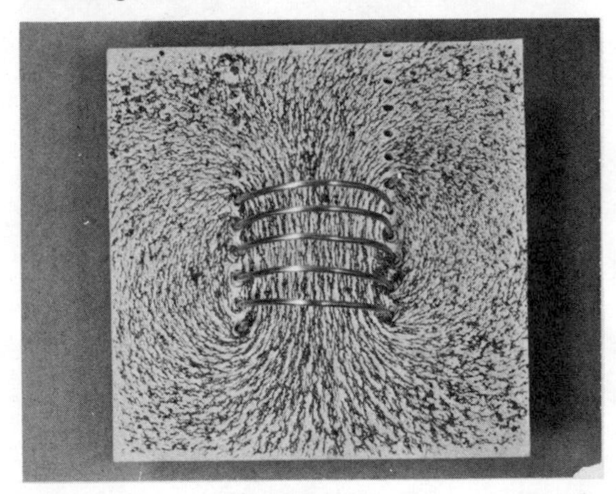

Figure 9-15 Iron-Filings Pattern of a Current-Carrying Coil

Such a coil is called an **electromagnet**, although this term is usually reserved for coils with an **iron core**. The electromagnetic force is concentrated in the centre of the coil and fans out at both ends. As in the case of a bar magnet, the flux lines form complete loops, and the magnetic field has an **N-pole** and an **S-pole**, as shown in Figure 9-16.

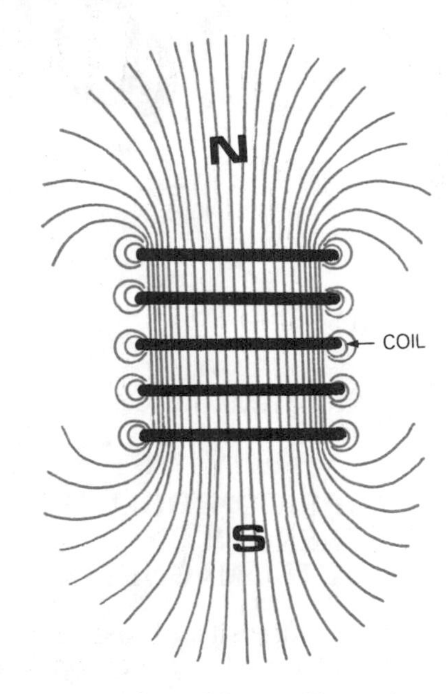

Figure 9-16 Lines-of-Force Map of a Current-Carrying Coil

If the direction of the electron flow through the coil is reversed, the magnetic poles reverse also.

To determine the **magnetic polarity** of a coil, the **left-hand rule for coils** is used:

> If a current-carrying coil is held in the left hand with the fingers pointing in the direction of the electron flow through the windings, the thumb will point toward the N-pole of the coil's magnetic field.

The electromagnetic field built up by the electron current **changes its intensity** whenever the current increases or decreases. When the electron current is interrupted, the magnetic field collapses and disappears.

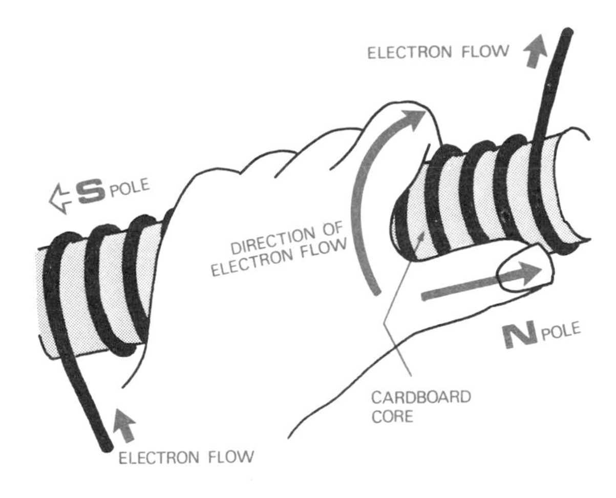

Figure 9-17 Left-Hand Rule for Current-Carrying Coils

9-5 FACTORS CONTROLLING THE STRENGTH OF AN ELECTROMAGNET

The strength or intensity of the electromagnetic field of a coil depends on **three factors**:

the number of windings of the coil
the strength of the electron current
the type of core material

The greater the **number of turns** in a coil, the more individual conductors will add their magnetic force to the total field, and the stronger will be the resulting electromagnetic force.

The greater the **electron current** flowing through a coil, the stronger will be the coil's field of force.

The higher the **permeability rating** of the metal used for the coil's core, the stronger will be the magnetic field created by the electron current.

The ability of an electron current to set up flux lines in the core of a coil is called **magnetomotive force**, symbol $\mathfrak{F}$. The magnetomotive force generated by a current-carrying coil depends on the number of turns in the coil **(N)**, and the electron current flowing through the coil **(I)**. The **international unit of magnetomotive force** is the **ampere-turn**, abbreviated **a.t.** One ampere-turn is the magnetomotive force produced by a current of one ampere flowing through a coil of one turn.

Thus, the magnetomotive force of a 300-turn coil carrying a current of 2.5 amperes is:

$$\mathfrak{F} = N \times I$$
$$\mathfrak{F} = 300 \times 2.5$$
$$\mathfrak{F} = 750 \text{ ampere-turns}$$

9-6 PRACTICAL ELECTROMAGNETS

The electromagnets used in motors, generators, meter movements, relays and other devices fall into two general groups: electromagnets for d.c. operation, and electromagnets for a.c.

A **d.c. electromagnet** has a core made of **solid** iron or soft steel (Fig. 2-3). The coil is wound on a special form called a bobbin. Its windings are usually protected by a layer of varnished tape.

Figure 9-18 A.C. Electromagnet

An **a.c. electromagnet** always has a **laminated** soft-steel core. Each sheet of soft steel is insulated from its neighbouring sheets by a thin layer of varnish or other insulating material. The lamination of the core prevents so-called **eddy currents** from being induced in the core by the continuously moving magnetic field of the alternating current. These eddy currents would generate heat and thus create a loss of energy in the core.

9-7 THE SOLENOID

Although the term solenoid may be used for any coil, it is generally applied only to coils having a **movable iron core**. This core or **plunger** travels inside the coil in a special, hollow insulator tube. The hollow tube is usually part of the coil bobbin.

Figure 9-19 A.C. Solenoid

Normally, the movable iron core is partially held outside the coil by a steel retaining spring. When the current is sent through the coil, **temporary magnetism** is induced in the iron core by the electromagnetic field. The mutual attraction between the electromagnetic force and the magnetism induced in the iron core pulls the core into the solenoid. When the current is switched off, the steel spring pulls the plunger back out.

The pulling force of a solenoid can be strong enough to open locks, operate electric switches, shift railroad switches, strike chimes, and perform many other mechanical operations. The solenoid is the **basic moving mechanism** used in electromechanical devices.

9-8 THE RELAY

A relay is an **electromagnetic switch** which is operated by remote control. It has an iron-core electromagnet with a movable iron armature hinged in front of the magnet's core. Switch contacts attached to the armature and the relay frame can be opened or closed by sending a small control current through the relay's coil.

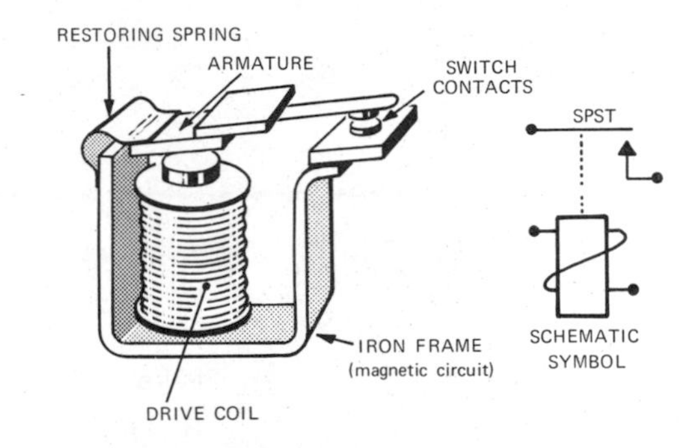

Figure 9-20 Basic Relay Design

Relays play a major role in telephone circuits, automated machinery and computers or wherever current and voltage levels must be changed by remote control.

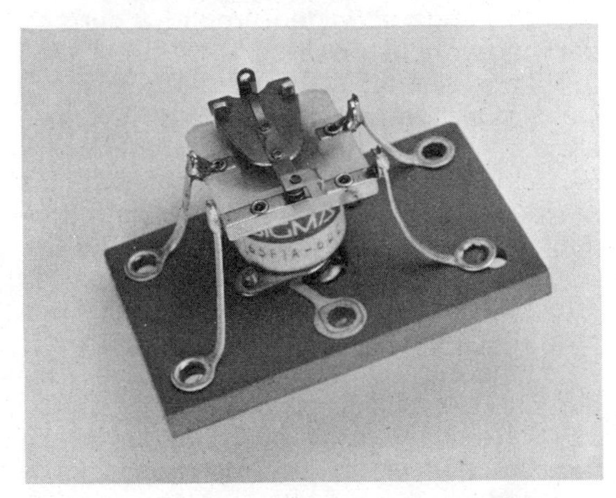

Figure 9-21 A Typical Relay

The starting circuit of an automobile is a typical example of a relay circuit. The heavy cables connecting the cranking motor to the battery must be as short as possible to avoid excessive energy loss in the cables' resistance. A heavy-duty relay, controlled by the starter switch, operates the cranking motor circuit when a small control current is sent through its coil.

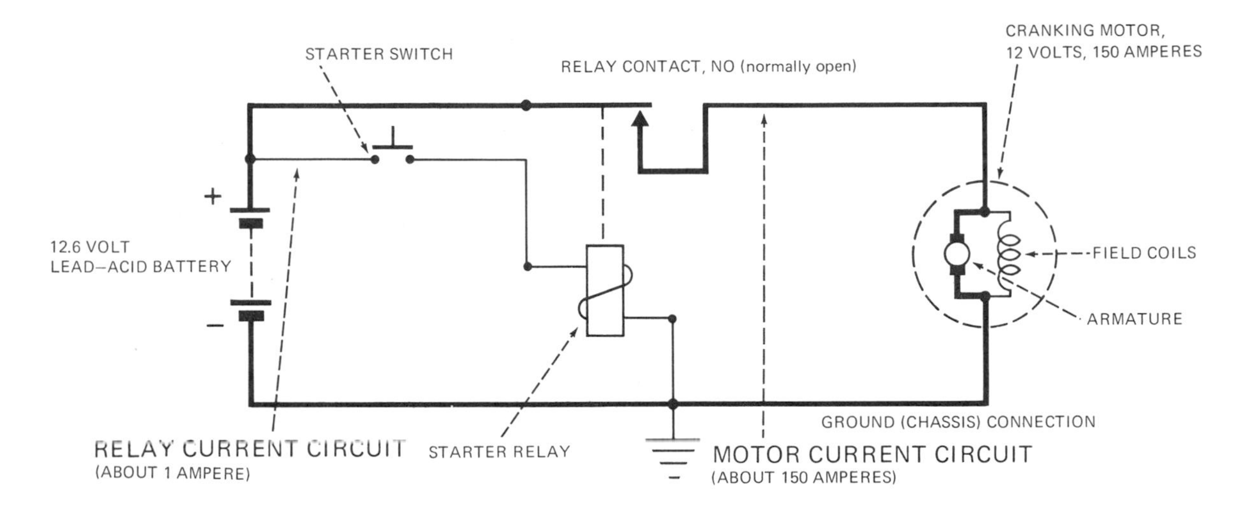

Figure 9-22 Cranking Motor Circuit

9-9 ASSIGNMENTS: ELECTROMAGNETISM

9-9A Write Full Answers

1. Describe Oersted's experiment.
2. Make a drawing to show the alignment of iron filings sprinkled around a current-carrying conductor.
3. State Oersted's principal discovery.
4. List all the characteristics of flux lines around a current-carrying conductor.
5. Explain how the direction of the lines of force around a current-carrying conductor can be shown experimentally.
6. Describe the relationship between the direction of electromagnetic lines of force and the electron current producing them.
7. State the left-hand rule for single conductors.
8. State the international rule on which the direction of an electromagnetic force is based.
9. Explain how parallel conductors carrying electron currents affect each other a) when the currents flow in the same direction b) in opposite directions.
10. Make drawings showing the shape of the flux lines around closely spaced conductors carrying a) currents flowing in the same direction b) in opposite directions.
11. Show how the flux lines around the individual turns of a coil merge to form a magnetic field.
12. State the left-hand rule for current-carrying coils.
13. Name the three factors that control the strength of an electromagnet and describe how each one contributes.
14. Define the term magnetomotive force and its unit of measurement, the ampere-turn.
15. Calculate the magnetomotive force generated by a 2,750-turn coil carrying a current of 0.25 amperes.

16. How do d.c. electromagnets differ from a.c. electromagnets?
17. What is the effect of eddy currents induced in the iron core of an electromagnet?
18. Use a simple, labelled drawing to explain how a solenoid works.

9-9B Indicate True or False

1. Oersted discovered that electromagnetism is caused by a conductor.
2. The electron current is the source of the electromagnetic force around a conductor.
3. Lines of electromagnetic force exist around a conductor at all times.
4. Since circular flux lines have no poles, magnetic compasses cannot be used to show their direction.
5. The direction of electromagnetic force depends on the electron current which creates it.
6. Closely spaced parallel conductors carrying electron currents always repel one another.
7. The flux lines around closely spaced conductors attract or repel, depending on the direction of the currents flowing through the conductors.
8. In a current-carrying coil, the flux lines of the individual conductors add their forces to form a strong electromagnetic field.
9. The flux lines generated by a current-carrying coil form complete loops enclosing the coil through its centre.
10. The strength of an electromagnet depends only on the number of its turns and the type of core.
11. The magnetomotive force of a coil is its ability to set up flux lines in its core.
12. The magnetomotive force generated by a coil depends on its core material.
13. A.c. and d.c. electromagnets differ in the construction of their cores.
14. A.c. electromagnets have laminated cores to reduce the amount of heat generated in them.
15. An electromagnet can be operated with a.c. or d.c., regardless of its core.
16. The plunger in a solenoid must be made of magnetic material.
17. The electromagnetic force developed by a solenoid is too weak to operate large electromechanical devices.

9-9C Select Correct Answer

1. Electromagnetism is caused by
   a) free electrons at rest
   b) free electrons flowing through a conductor
   c) a conductor
   d) any electric charge
2. Lines of electromagnetic force around a conductor
   a) exist only as long as a current flows through the conductor
   b) are independent of current flow
   c) move along the conductor
   d) form various patterns around the conductor
3. Flux lines around a current-carrying conductor
   a) have no specific shape or direction
   b) act only at specific spots along the conductor
   c) form a circular, concentric pattern with the conductor at the centre
   d) have an N-pole and an S-pole
4. The direction of electromagnetic lines of force
   a) is independent of all other factors
   b) depends on the direction of the electron current
   c) cannot be determined because they are circular
   d) does not depend on the electron flow
5. Flux lines around a current-carrying conductor
   a) act in any direction
   b) have no particular direction
   c) change their direction along the conductor
   d) act in the direction in which they push an N-pole

6. The left-hand rule for single conductors
   a) relates the direction of the flux lines to the direction of the electron current
   b) does not depend on the direction of the electron current
   c) is of no practical use
   d) is used to determine the direction of the electron current flowing through a conductor
7. If a magnetic compass is brought into the electromagnetic field around a conductor,
   a) its needle is aligned at right angles to the direction of the flux lines
   b) its needle aligns itself along the conductor
   c) its needle is aligned along the paths of the flux lines
   d) the flux lines do not affect the compass
8. In a current-carrying coil,
   a) the flux lines of the individual windings tend to cancel each other
   b) the flux lines of the individual turns merge into a magnetic field similar to that of a bar magnet
   c) the flux lines are concentrated in the space around the coil
   d) the polarity of its magnetic field is independent of the electron current
9. A current-carrying coil
   a) bears no relationship to an electromagnet
   b) does not have magnetic poles
   c) sets up flux lines only inside the coil
   d) is an electromagnet without a core
10. The left-hand rule for current-carrying coils
    a) is equally applicable to single conductors
    b) relates the direction of the flux lines to the direction of the electron current
    c) shows the direction of electron flow
    d) can be used even when it is not known how the coil is wound
11. The intensity of the magnetic field of an electromagnet
    a) depends on the type of core material and the amount of electron current
    b) is independent of the core material
    c) depends on the number of turns, the current in amperes, and the type of core
    d) is independent of the number of windings in its coil
12. Magnetomotive force
    a) is measured in ampere-turns and is independent of the core material
    b) refers to the number of flux lines generated in the core of a current-carrying coil
    c) is not influenced by the number of turns in a coil
    d) is independent of the electron current flowing through the coil
13. The core of an a.c. electromagnet is laminated
    a) to increase its mechanical stability
    b) because heat is more readily dispersed
    c) to distinguish it from a d.c. electromagnet
    d) to prevent eddy currents from being induced in it
14. The plunger of a solenoid
    a) is a tubular air core
    b) depends on induced, temporary magnetism for its operation
    c) can be made of any metal
    d) returns to its normal position without assistance
15. Solenoids
    a) are electromagnets without a core
    b) are used in electric motors and generators
    c) are electromechanical devices that exert a pulling or pushing force
    d) have a fixed soft-steel core

10

ELECTRIC ENERGY AND POWER

The main purpose of any electric circuit is to put electricity to work carrying energy from a supply to a load. The **amount of energy transported** to a load depends on **three factors**:

- the **voltage** acting on the free electrons
- the amount of **electron current** flowing in the circuit
- the length of **time** the electrons are at work

The **international unit of measurement** for electric **energy** is the **joule**, also called the **watt-second**. One joule is the amount of energy carried by an electron current of one ampere flowing for a period of one second with a pressure of one volt.

The joule is too small for practical purposes. So much electric energy is used even in an average home that a much larger unit, the **kilowatt-hour**, is used to keep track of it. One kilowatt-hour, or kWh, equals 3,600,000 joules. (The metric prefix **kilo** stands for 1,000 and one watt-hour equals 3,600 watt-seconds.)

10-1 THE KILOWATT-HOUR METER

A small electromechanical computer, the kilowatt-hour meter, is used to measure the electric energy consumed in a home or shop. This instrument performs **several functions at the same time**: it senses **voltage** and **electron current**, measures elapsed **time**, multiplies the three quantities automatically, and continuously records the number of consumed kilowatt-hours on its circular dials.

Figure 10-1 Kilowatt-hour Meter

A kilowatt-hour meter begins to register the energy consumed in a home at the moment the building is first connected to the power distribution system. From then

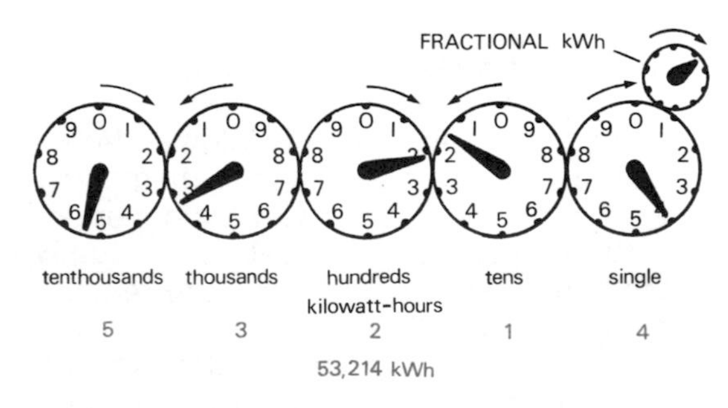

Figure 10-2 Indicator Dials of a Kilowatt-hour Meter

on, its pointers keep on totalling kilowatt-hours for decades. From time to time, usually at two-month intervals, its dials are read to determine the number of kilowatt-hours consumed during that period.

How to read a kilowatt-hour meter: the pointers of the kilowatt-hour meter rotate in **opposite directions**, as shown in Figure 10-2.

From left to right, the circular dials indicate ten-thousands, thousands, hundreds, tens and single kilowatt-hours. The pointer positions are read by noting the number at which the indicator is pointing or has just passed.

Calculating an energy bill: Figures 10-3 and 10-4 show the same kWh meter in April and in June of the same year.

Figure 10-3 kWh Meter—April Reading

Figure 10-4 kWh Meter—June Reading

Read the number of kilowatt-hours indicated by the meter at the beginning and at the end of the two-month period. Find the number of kilowatt-hours **actually consumed** by subtracting the April reading from the June reading.

The cost of this energy depends on the **rate structure** of your local power company. Usually, the rate per kilowatt-hour (kWh) is highest for the first 50 or 100 kWh, drops to a lower price for the next 200 kWh, and so forth. Consult your local hydro office for detailed information.

10-2 ELECTRIC POWER

Any electric device has three important ratings: voltage, current, and power. The power rating indicates how much energy the device will use in each second.

The **international (S.I.) unit of power** is the **watt**, symbol **W**. One watt equals an energy consumption of one joule per second. A device has a power rating of one watt if it draws a current of one ampere from a one-volt source for a period of one second.

Power is calculated by multiplying applied voltage and electron current:

where P is power in watts
E is voltage in volts
and I is current in amperes

Figure 10-5 General Equation for Power

Here is a **sample calculation** of power: a small incandescent lamp draws a current of 0.25 amperes from a 120-volt supply. What is its power rating?

1. $P = E \times I$
2. $P = 120 \times 0.25$
3. **$P = 25$ watts**

In practical work, power is measured with a **wattmeter.**

10-3 THE WATTMETER

Most wattmeters come equipped with a linecord and a built-in receptacle to plug in the device to be tested.

Figure 10-6 Wattmeter

Here is how the wattmeter is used: plug the meter's linecord into a 115-volt wall outlet. Select the highest wattage range on its range switch. Plug the linecord of the appliance to be tested into the meter's own receptacle. Turn the appliance on and read its power rating on the meter scale. Switch to a lower range if the pointer does not move too far on the scale.

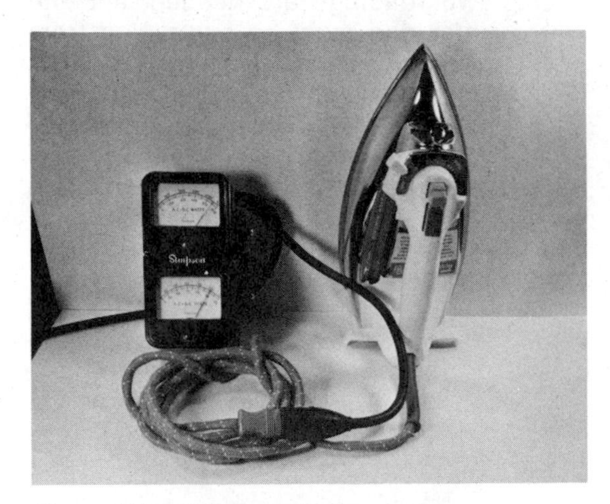
Figure 10-7 Wattmeter in Use

The wattmeter is an important service tool. A check of the actual power consumption of an appliance will quickly show the condition of the device and eliminate much guesswork in its repair.

10-4 ASSIGNMENTS: ELECTRIC ENERGY AND POWER

10-4A Write Full Answers

1. State the three factors that determine the amount of electric energy carried by free electrons.
2. Define the international unit of energy, the joule.
3. Define the practical unit of electric energy, the kilowatt-hour.
4. Describe the several operations performed automatically by a kilowatt-hour meter.
5. Explain why a kilowatt-hour is equal to 3,600,000 joules, or watt-seconds.
6. Make a neat drawing of the dials of a modern kilowatt-hour meter. Indicate the direction of rotation of its pointers and the kWh value of each dial.
7. Calculate the bill for the energy consumption indicated in Figures 10-4 and 10-5, assuming a charge of 4 cents for the first 50 kWh, 3 cents for the next 200 kWh, and 2 cents for the remaining kilowatt-hour.
8. Phone your local hydro office and find out their current residential rates for electric energy.
9. Define the term power.
10. Define the international unit of power, the watt, in two ways.
11. Calculate the power rating of the following devices, using proper mathematical procedure:
   a) a hand drill drawing 6 amperes from a 115-volt source
   b) a TV set operating with 1.8 A and 110 V
   c) a cranking motor drawing 120 A from a 12-volt battery
   d) a transistor radio using 0.020 A and 9 V
   e) a solenoid drawing 0.32 A from a 24-volt supply
12. Describe how a modern wattmeter is used for testing the power rating of a small appliance.

10-4B Indicate True or False

1. The greater both voltage and current

flow, the more energy will be delivered to a load during a given time.
2. The kilowatt-hour is a much smaller unit of energy than the watt-hour.
3. The practical unit of energy is equal to 1,000 watt-hours.
4. A kilowatt-hour meter measures voltage and current to arrive at the total consumed energy in kilowatt-hours.
5. The dials of a kilowatt-hour meter all have the same kWh value.
6. To calculate an energy bill, the earlier meter reading must be subtracted from the later one.
7. The international unit of power, the watt, is equal to an energy consumption of one joule per second.
8. The lower the voltage rating of an appliance, the more power it will use.
9. A wattmeter automatically measures voltage and current, multiplies the two electrically and indicates power directly in watts.
10. The watt and the watt-second are the same unit of measurement.

10-4C **Select Correct Answer**

1. The energy delivered to a load depends on
   a) the amount of electron current and length of time
   b) the voltage acting on the electrons
   c) applied voltage, electron current, and length of time
   d) the electron current and the applied voltage
2. In the measurement of electric energy
   a) time is unimportant
   b) elapsed time is an important consideration
   c) only voltage and current are important
   d) the resistance of the load is of prime importance
3. One kilowatt-hour is equal to
   a) 3,600 watt-hours
   b) 100 watt-hours
   c) 3,600 joules
   d) 3,600,000 joules
4. The kilowatt-hour meter
   a) senses voltage and current and measures time
   b) must be read from right to left
   c) has pointers rotating in the same direction
   d) always returns to zero after each reading has been taken
5. In calculating the electric energy consumed in a home,
   a) a single reading of the kilowatt-hour meter is sufficient
   b) two readings are needed, and the later one must be subtracted from the earlier one
   c) two readings are used, and the earlier one must be subtracted from the later one
   d) the interval between the two readings is of no importance

11
ELECTRIC LAMPS AND HEATING ELEMENTS

The **heating effect** was the first characteristic of the electron current to be put to large-scale use. When free electrons are forced through a conductor having a high internal resistance, their kinetic energy is converted to heat energy. The higher the voltage and resistance of the conductor, the more heat will be generated. Eventually, the conductor's temperature will become so high that the metal will melt and vapourize. If the oxygen of the air is kept away from it, the incandescent metal will give off a bright light.

Incandescent lamps and electric heating elements use this heating effect in their operation. Since most of the electric energy is turned into heat, incandescent lamps are very inefficient light sources.

Glow lamps and fluorescent lamps do not use the heating effect of a current to generate light. In these devices, electron energy is converted directly to light energy, although much energy is lost in the conversion process.

11-1 INCANDESCENT LAMPS

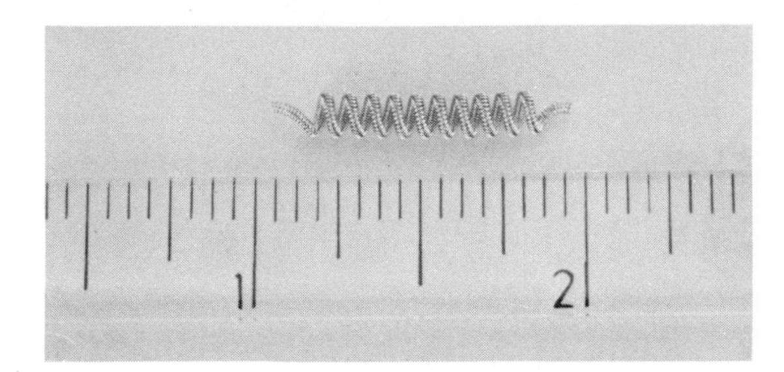

Figure 11-1 Tungsten Filament (Enlarged)

The main part of an incandescent lamp is its **filament**, a densely wound coil of **thin tungsten** wire.

Tungsten is used because it is the only metal which retains its shape even when white hot (incandescent).

The typical design of incandescent lamps for home use is shown in Figure 11-2. During manufacture, the air is removed from the glass bulb, and an inert gas such as argon or nitrogen is introduced. This gas prevents tungsten atoms from being boiled off the filament, thus increasing the life of the lamp.

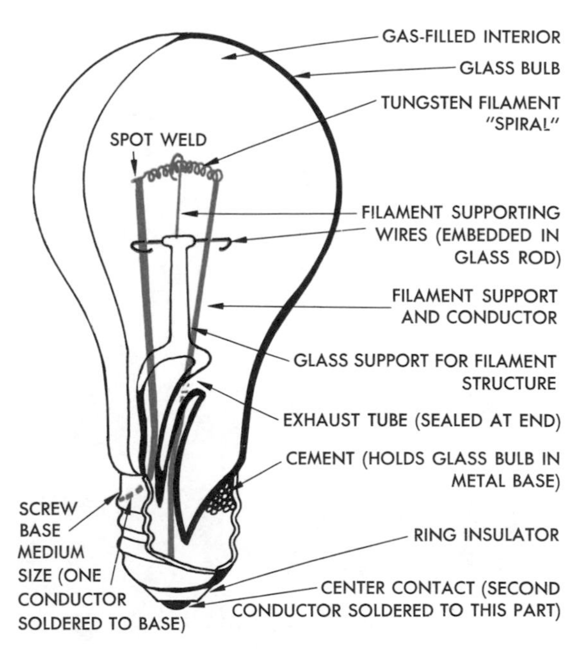

Figure 11-2 Incandescent Lamp

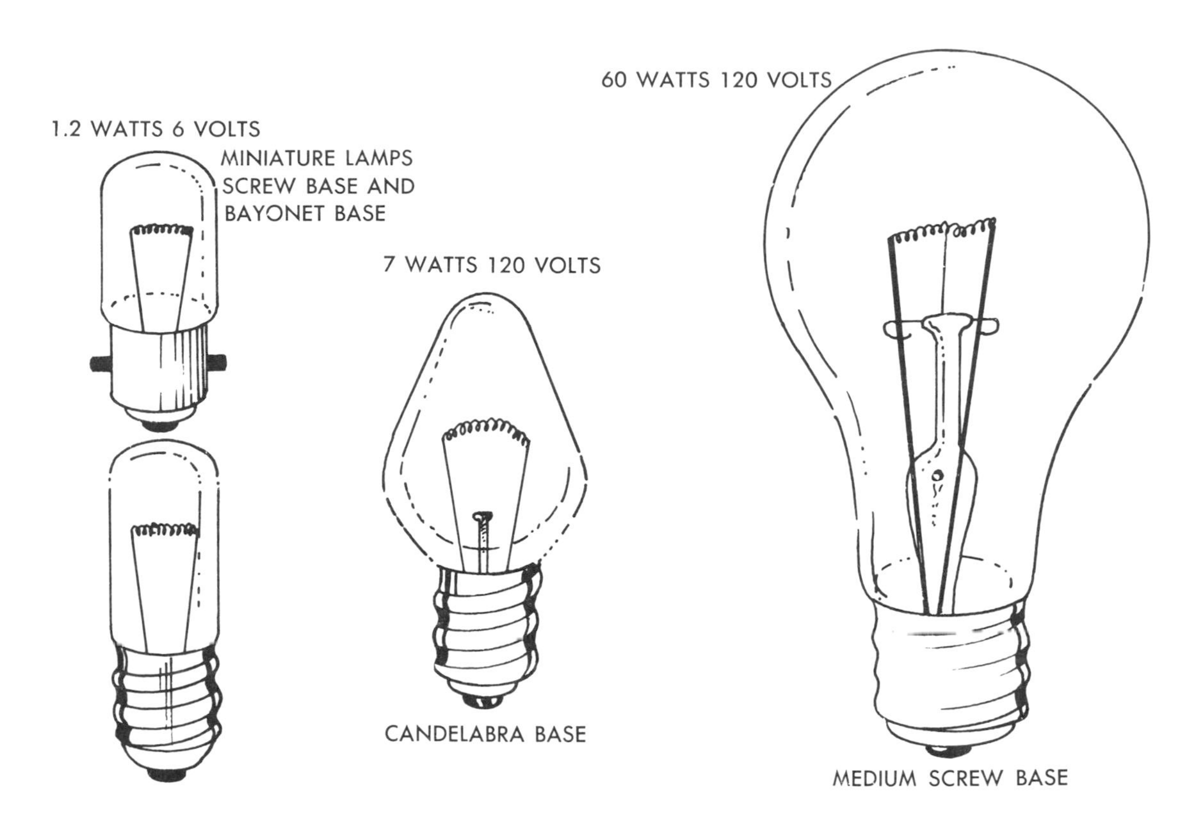

Figure 11-3 Various Lamp Sizes and Bases

Incandescent lamps are available in many sizes and voltage and power ratings and with a variety of bases. The **standard medium base** is the most common incandescent lamp base.

A recent development is the **quartz-iodine incandescent lamp**. This lamp also has a tungsten filament, but its envelope is made of **quartz** and is filled with **iodine gas**. Since quartz can stand much higher temperatures than glass, these lamps are of **much smaller size** than normal incandescent lamps with the same power rating.

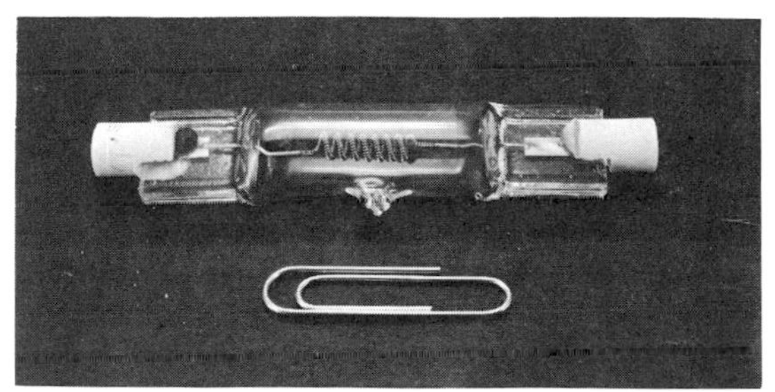

Figure 11-4 Quartz-Iodine Incandescent Lamp

Caution: quartz reacts with skin oils at high temperatures. Always use a paper napkin or cloth when handling quartz-iodine lamps, or their life will be drastically shortened.

11-2 GLOW LAMPS

Glow lamps have **no filament**. Their glass envelope is filled with a special gas, usually **neon**, at low pressure. This gas is **ionized**, or made to glow, by voltages in excess of 60 volts.

When a sufficiently high voltage acts on the gas atoms between the lamp's electrodes, the following actions take

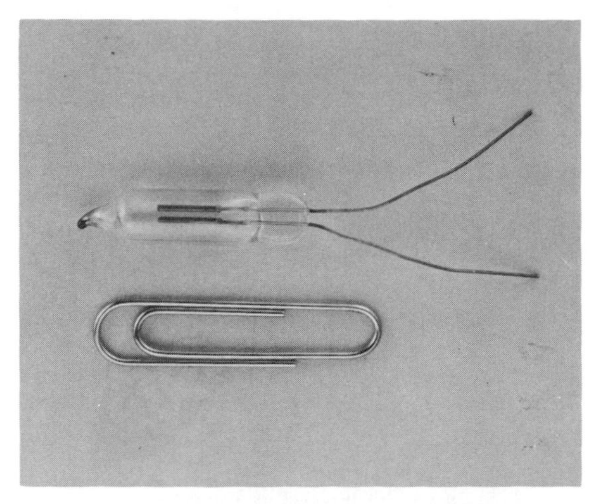

Figure 11-5 Neon Glow Lamp, Type NE-2

place: electrons are pulled out of the gas atoms by the positively charged rod, creating **positive gas ions**. These ions are attracted by the negatively charged electrode and drift toward it. At the negative electrode, excess electrons jump on the positive gas ions and **neutralize** them. Each electron jumping across **gives off energy** as a brief flash of light, called a **photon**. Since thousands of photons are generated at every instant, the gas around the **negative electrode** glows brightly. The colour of this light depends on the kind of gas with which the lamp is filled.

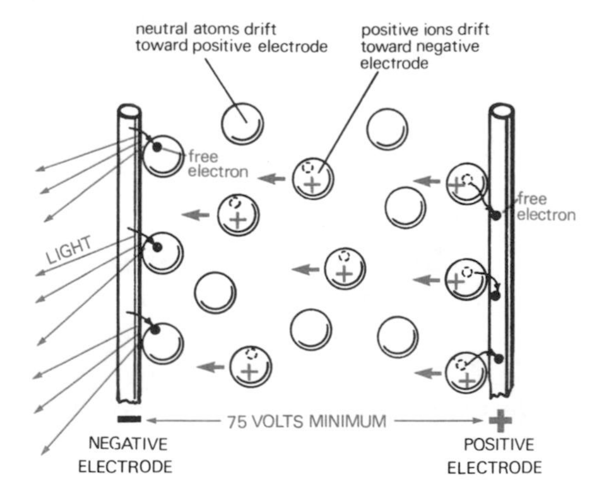

Figure 11-6 Inside the Glow Lamp

If a glow lamp is connected to an a.c. supply, both electrodes will glow due to the rapidly alternating polarity of the alternating current.

For technical reasons, glow lamps must always be connected in series with a **current limiting resistor**, or they will burn out. This resistor is usually concealed in the base of the lamp.

11-3 ELECTRIC HEATING ELEMENTS

Electric appliances that depend on heat for their operation use so-called **heating elements** to convert electricity to heat energy. These elements are available in many sizes and shapes, with power ratings ranging from a few watts to several kilowatts.

Figure 11-7 Typical Heating Element

The main part of a heating element is the **resistance wire**, or resistor. It is made of nichrome, constantan, alloy 800, or some other high-resistance alloy.

Toaster heating elements, for example, consist of ribbon-type resistance wire wound on sheet mica. The mica supports the resistor and insulates it from the metal frame of the toaster.

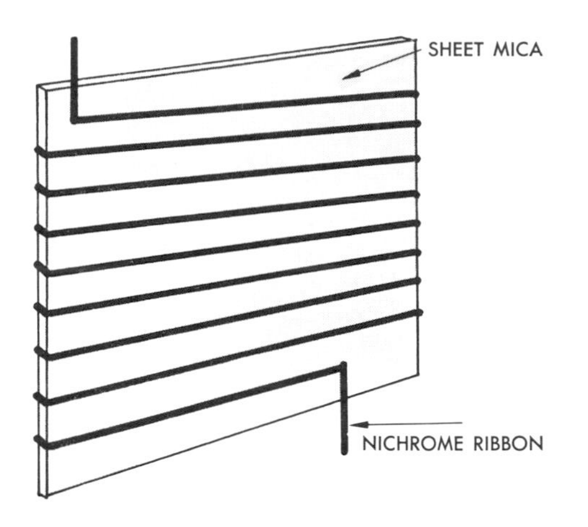

Figure 11-8 Heating Element for a Toaster

A common heating element is the tubular heater used in kitchen ranges, baseboard heaters, electric kettles and other appliances. Its resistance wire is totally enclosed in a steel tube, and the space between the resistor and the tube is filled with a heat-conducting ceramic insulator.

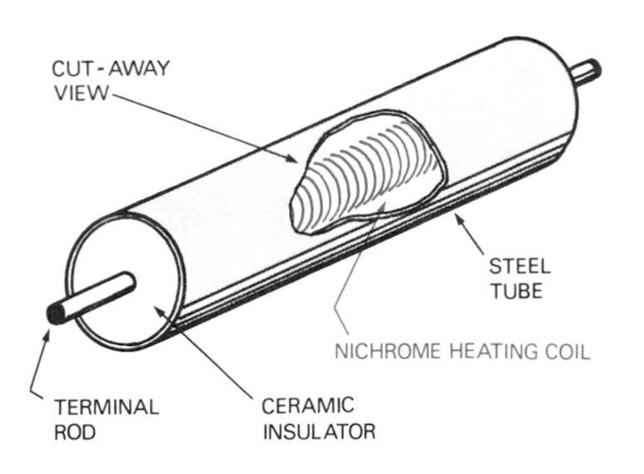

Figure 11-9 Tubular Heating Element

The tubular heating element of a baseboard heater is studded with hundreds of aluminum fins to provide a large contact surface for the air to be heated.

Figure 11-10 Electric Baseboard Heater

The heating element of the soldering iron shown in Figure 11-11 surrounds the shaft of the heating tip. A heat-conducting insulator provides such efficient heat transfer that only 47 watts are required to heat the tip to operating temperature.

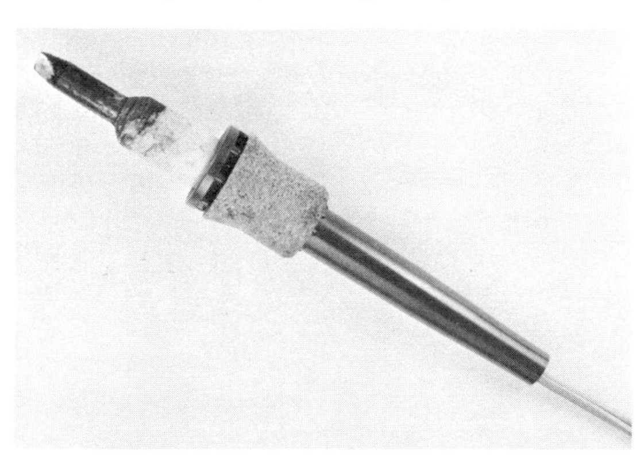

Figure 11-11 Electric Soldering Iron

11-4 ASSIGNMENTS: ELECTRIC LAMPS AND HEATING ELEMENTS

11-4A Write Full Answers

1. Describe the so-called heating effect of an electron current.
2. What will happen when a metal is heated to incandescence in open air? Why?

3. Make a labelled drawing showing the main parts of an incandescent lamp.
4. State the purpose of the inert gas contained in a modern incandescent lamp.
5. Describe how the incandescent lamp converts electric energy to light energy.
6. Define the words incandescent and incandescence.
7. Name the main types of lamp bases used for incandescent lamps.
8. Make a labelled drawing of a quartz-iodine lamp.
9. Why must a quartz-iodine lamp not be handled with bare hands?
10. Draw a sketch showing the internal construction of a typical neon lamp. Label all parts.
11. Explain how a glow lamp generates light around its negative electrode.
12. Why does a glow lamp connected to an a.c. source generate light at both electrodes?
13. Describe the internal construction of a heating element for a toaster. Make a labelled drawing to illustrate your answer.
14. Describe the internal construction of a tubular heating element.

11-4B **Indicate True or False**

1. A wire heated to incandescence by an electric current will produce usable light.
2. The higher the applied voltage and resistance of a conductor, the more heat will be generated in it.
3. A conductor heated to incandescence in air will burn up instantly.
4. No energy is lost in the conversion of electricity to light in the incandescent lamp.
5. A quartz-iodine lamp is of smaller size and weight than a standard lamp due to the shape of its filament.
6. The quartz envelope of a quartz-iodine lamp must not be touched with bare hands.
7. Glow lamps use the heating effect of an electron current to light up the gas.
8. In a glow lamp, only the negative electrode produces light.
9. Both electrodes of a glow lamp will generate light when the lamp is connected to a d.c. supply.
10. A glow lamp must be connected in series with a current-limiting resistor.
11. The main part of a heating element is its resistance wire, or resistor.
12. The purpose of the steel tube of a tubular heater is to conduct electricity to the resistance wire.

11-4C **Select Correct Answer**

1. If the electron current flowing through a conductor becomes excessive,
   a) the conductor will become warm
   b) the electron-atom collisions inside the conductor will create little heat
   c) the conductor will not be affected
   d) the conductor will heat up, become incandescent, and burn up
2. The greater the electron current flowing through a resistance,
   a) the less effect the current will have on the conductor
   b) the more electron-atom collisions occur, and the more the heat will be generated
   c) the less effect the resistance will have on the electrons
   d) the less heat will be produced
3. A conductor is heated to incandescence
   a) when it glows with a deep red colour
   b) when it gives off heat, but has not changed its colour
   c) when it gives off a brilliant white light
   d) when it glows with a bright orange light
4. A metal heated to incandescence in air
   a) is not affected at all
   b) will glow brightly as long as the current is flowing
   c) can be used as a light source
   d) reacts with the oxygen and burns up
5. An incandescent lamp
   a) converts electricity to light very efficiently

b) usually has a quartz bulb
c) is extremely inefficient, since most of the electric energy is lost as heat
d) uses the inert gas to convert electricity to light energy

6. In a glow lamp
   a) the electrodes give off light
   b) both electrodes generate light energy
   c) the gas is ionized if the applied voltage exceeds a critical value
   d) the gas is not affected by the applied voltage
7. The voltage applied to a glow lamp
   a) is of no importance in the operation of the lamp
   b) will fire the lamp with as little as 12 volts
   c) must be in excess of 60 volts to ionize the gas
   d) does not affect the gas at all
8. A glow lamp connected to an a.c. source
   a) will not work at all
   b) will generate light at both electrodes
   c) will produce light at one electrode only
   d) will fire below 60 volts
9. In a heating element, the insulator surrounding the resistance wire
   a) can be made of any insulating material
   b) is usually a thermoplastic material
   c) is a heat-conducting ceramic insulator
   d) does not influence the operation of the element

12
CIRCUIT PROTECTION DEVICES

Electric circuits are designed to operate with a **specific voltage** and a **maximum allowable current** rating. If either or both of these limits are exceeded, a circuit may be damaged or even destroyed.

The three possible causes for dangerous levels of voltage or current are **overload, short circuit** and **lightning**. To protect circuits from the potentially destructive effects of these conditions, circuit protection devices are used.

12-1 OVERLOAD, SHORT CIRCUIT AND LIGHTNING

Figure 12-1 Overloaded Receptacle

An **overload** occurs when a circuit carries more current than it can safely handle. If you plug a toaster, a TV set and an electric heater into the same receptacle, that circuit will be **dangerously overloaded**. If the circuit is protected with a fuse of the correct size, generally **15A**, the fuse link will melt, break the circuit, and remove the danger.

Figure 12-2 Fuse Link Melted by Overload Current

Figure 12-3 Fuse Link Vapourized by Short-Circuit Current

If a fuse with too high a current rating (25A or 30A) is used for the receptacle circuit, the conductors will heat up and their insulation will melt. If the overload continues, a **short circuit** will occur and blow out the fuse link, as shown in Figure 12-3.

Unfortunately, in most cases of severe and prolonged overload, a **fire** will start long before the short circuit finally blows the over-rated fuse.

Overloads can be prevented if only **one appliance** is plugged into a receptacle, and if all common receptacle circuits are protected by **15-ampere fuses.**

Figure 12-4 Typical Home Fuse Panel

A **short circuit** is a path of extremely low resistance which allows the circuit current to **bypass the load**. If a linecord is frequently twisted, the insulation between the conductors will break and the wires will touch, causing a short circuit.

At that instant, the electrons will **bypass** the normal load resistance and flow through the short circuit. This drastic reduction in circuit resistance causes a huge increase in current flow. If the outlet is protected by a 15A fuse, the fuse link will be vapourized almost instantly, and no further damage can occur.

But if **too large a fuse** is used in the receptacle circuit, the short-circuit current will melt the linecord, splash around white-hot metal and usually **cause a fire**.

Short circuits can be avoided by frequently checking the condition of linecords and by having defective cords and appliances repaired.

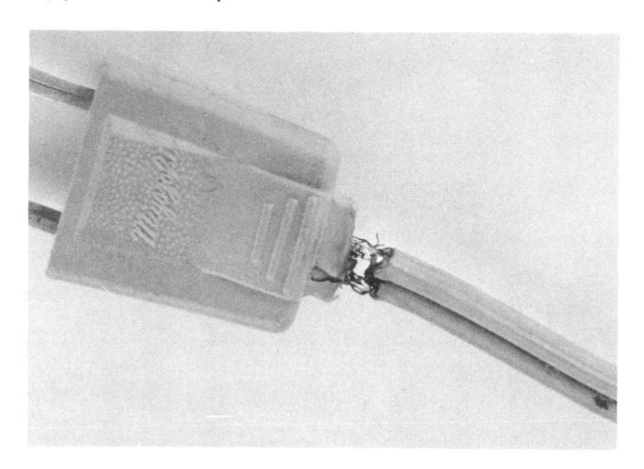

Figure 12-5 Short Circuit in Faulty Linecord

Lightning damage is comparatively rare. If a lightning bolt strikes an electric circuit, it will vapourize the conductors, melt the equipment, and often set fire to the building. Fortunately, such damage can be avoided by using **lightning rods**

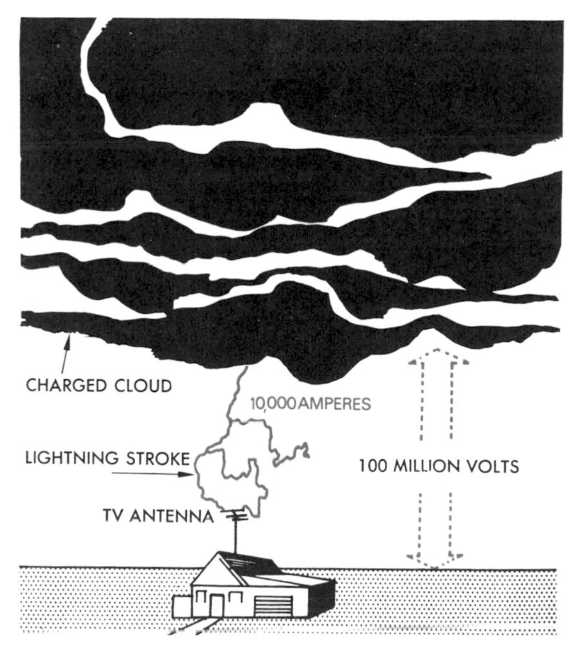

Figure 12-6 Voltage and Current Levels of a Typical Lightning Bolt

and **arrestors** to protect both building and circuits.

Scientific measurements of lightning strokes place the voltage of an average bolt at over 100,000,000 volts with an average discharge current of 10,000 amperes. The only thing we can do with these staggering amounts of electricity is **drain them into the earth**.

12-2 FUSES

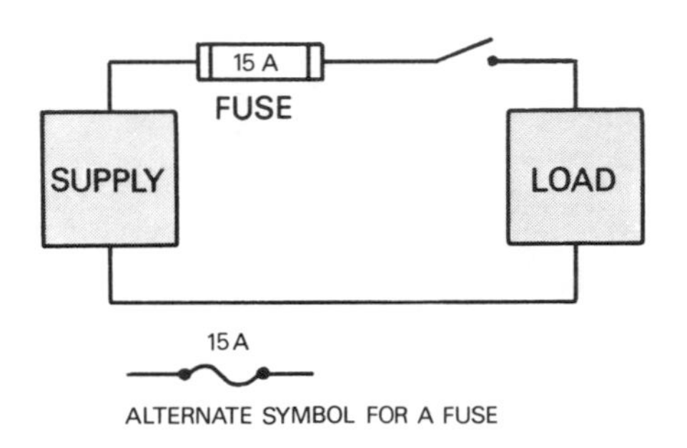

Figure 12-7 Typical Series Connection of a Fuse

A fuse is always connected **in series** with the device or circuit it protects. This forces the entire circuit current to flow **through the fuse link**, a thin strip of fusible metal that will melt when the current exceeds a critical value, the fuse rating.

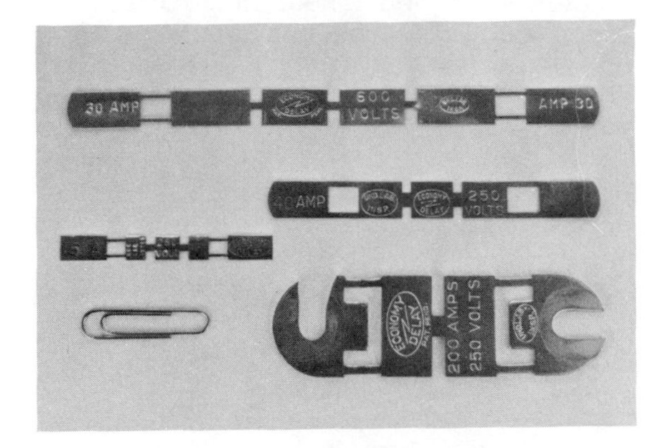

Figure 12-8 Various Fuse Links

The higher the blow-out current rating of a fuse, the larger is its physical size. Fuses range from the 0.1A miniature cartridge fuses used in electronic equipment, all the way to the large 600A knife-blade fuses used in power distribution circuits.

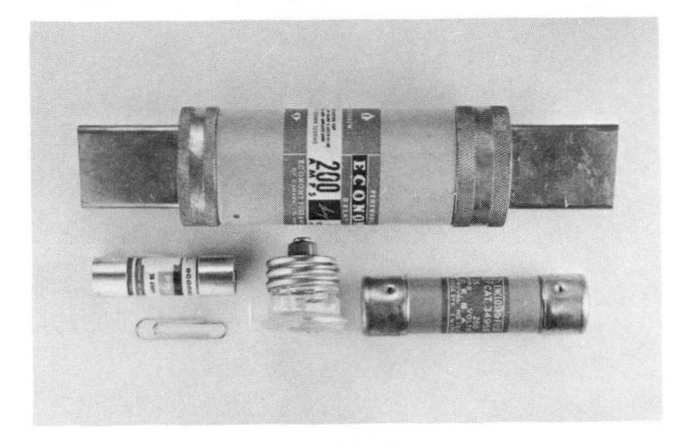

Figure 12-9 Some Typical Fuses

12-2A Miniature Cartridge Fuses

These fuses consist of a glass cylinder with metal ferrules at both ends. The fuse link is a thin wire of fusible alloy, as shown in Figure 12-10. Miniature cartridge fuses are used chiefly in low-power electronic equipment.

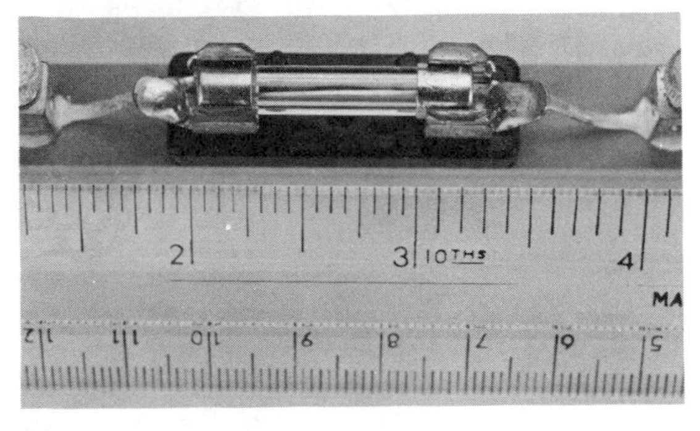

Figure 12-10 Miniature Cartridge Fuse in Fuse Holder

These fuses are available with current ratings from 1/100A to 30A, and voltage ratings from 32V to 250V.

12-2B Plug Fuses

These are the common fuses used in home circuits. The plug fuse has a glass casing through which the fuse link can be inspected. The current rating is shown next to the fuse link and is also stamped on the base contact of the fuse.

Figure 12-11 Plug Fuse

Plug fuses have standardized current ratings of 10A, 15A, 20A, 25A and 30A. Their common voltage rating is 125V.

12-2C **Cartridge Fuses**

Cartridge fuses are manufactured in two basic types: disposable and renewable fuses. Disposable cartridge fuses find their principal use in major appliance circuits, such as kitchen ranges and electric dryers.

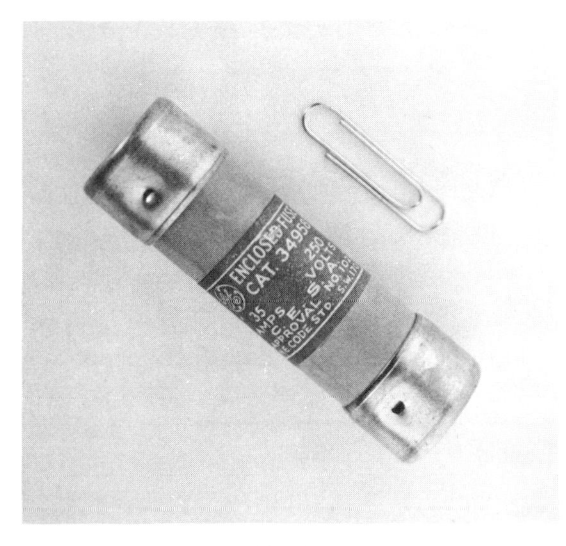

Figure 12-12 Disposable Cartridge Fuse

Renewable cartridge fuses have a replaceable fuse link. This type of fuse is used mainly in industrial circuits where blow-outs and overloads occur more frequently than in the home.

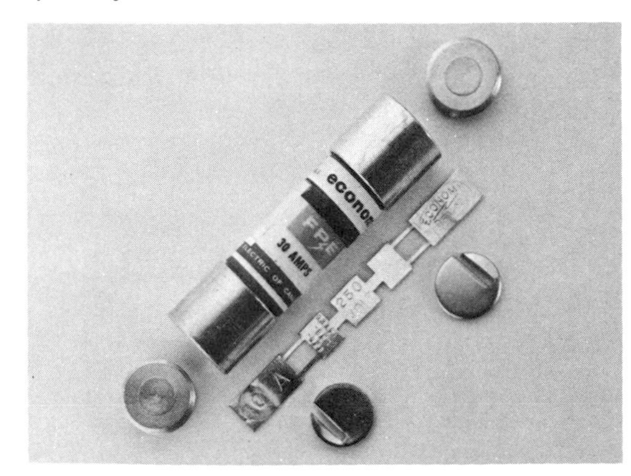

Figure 12-13 Renewable Cartridge Fuse — Disassembled

Cartridge fuses have current ratings ranging from 30A to 60A, and voltage ratings from 250V to 600V.

12-2D **Knife-Blade Fuses**

Knife-blade fuses are cartridge fuses for high-power industrial circuits. Some of these are huge, often weighing several pounds each. Most knife-blade fuses are of the renewable type, especially the larger ones.

Figure 12-14 Renewable Knife-Blade Fuse — Disassembled

Knife-blade fuses have current ratings ranging from 60A to 600A, with a maximum voltage of 600V.

12-3 CIRCUIT BREAKERS

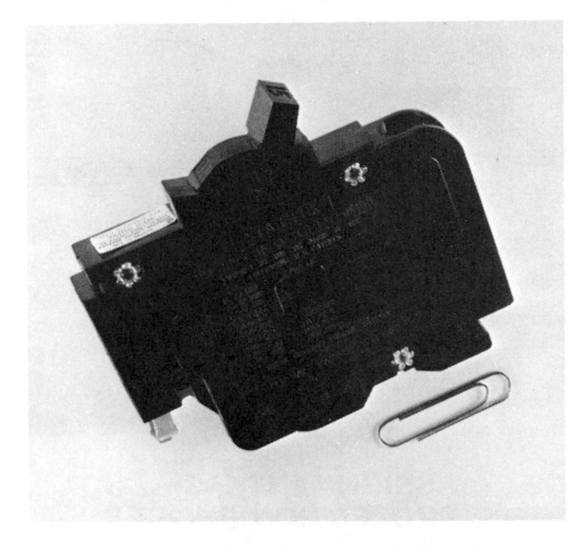

Figure 12-15 Typical Circuit Breaker, 15A

A circuit breaker is a **thermal** or **electromechanical switch** that breaks the current flow in case of overload or short circuit. It can be reset hundreds of times by simply flicking its lever back into the "on" position.

Figure 12-16 Bimetallic Strip Circuit-Breaker Mechanism

Circuit breakers are made in three basic types: thermal breakers, electromagnetic breakers, and thermal-electromagnetic breakers.

The **thermally actuated circuit breaker** has a **bimetallic strip** as its current-sensing element. Such a strip is made of two unlike metals welded together. When the current flowing through the strip exceeds its maximum allowable level, the heat generated by it will **bend the strip**, snap open the contacts, and break the circuit.

The thermally actuated circuit breaker reacts rather slowly to overload conditions, but breaks a circuit quickly in case of a short circuit.

The **electromagnetic circuit breaker** uses an electromagnet as its current-sensing mechanism. When the current flowing through the electromagnet exceeds the maximum allowable value, the magnetic force generated by it opens the breaker contacts and breaks the circuit.

12-4 LIGHTNING RODS AND ARRESTORS

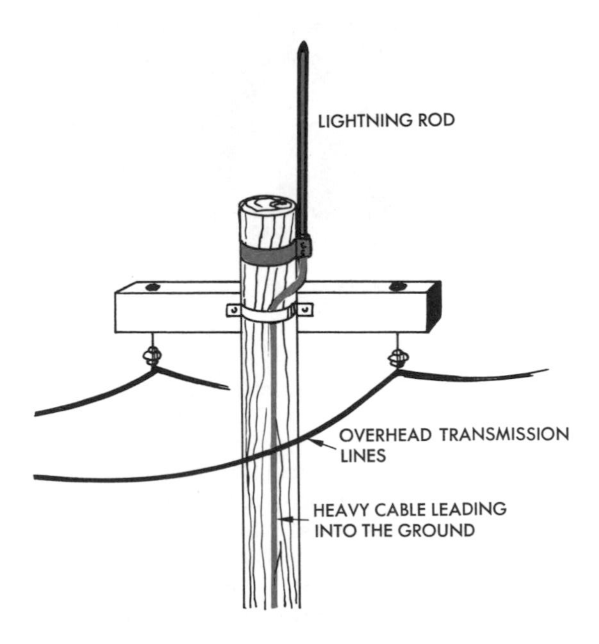

Figure 12-17 Lightning Rod Installation

A **lightning rod** provides a highly conductive path into the earth for atmospheric electricity. Placed at the highest points of the structure to be protected, lightning rods are connected to the ground by heavy copper conductors. The down path must be as short as possible, and the ground terminal must make excellent contact with the soil.

Buildings protected by lightning rods are virtually lightning proof. Lightning rods often discharge atmospheric electricity before a lightning stroke can occur.

Lightning arresters protect electric circuits from damage caused by atmospheric electricity. They are generally installed at the point where overhead wires, such as TV antenna leads and telephone lines, enter a building. Their purpose is to provide a **direct ground path** for atmospheric electricity without affecting the normal signal currents carried by these lines.

Figure 12-18 Lightning Arrester for TV Antenna (Mounted on Cold-Water Pipe)

Lightning arresters are usually mounted directly on a **cold-water pipe**, or connected to the nearest cold-water pipe by a heavy copper conductor.

The main part of a lightning arrester is the **flash-over gaps** concealed in its base. If a high voltage surge occurs in the signal line connected to the arrester, an **arc** will be formed across the gaps and the excess energy will flow into the earth, bypassing the equipment connected to the signal wires.

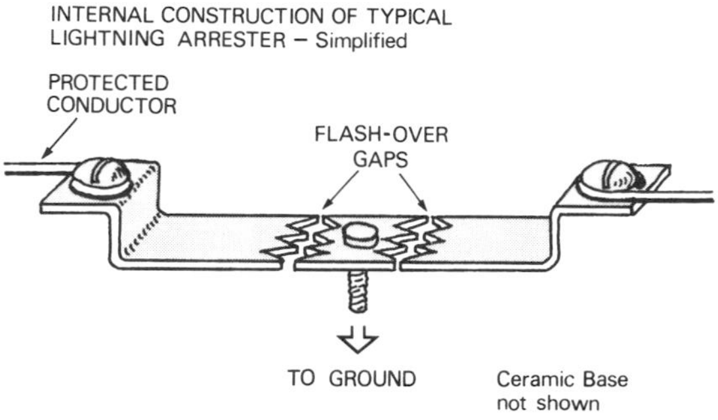

INTERNAL CONSTRUCTION OF TYPICAL LIGHTNING ARRESTER – Simplified

PROTECTED CONDUCTOR

FLASH-OVER GAPS

TO GROUND

Ceramic Base not shown

Figure 12-19 Flash-Over Gaps in Lightning Arrester

12-5 ASSIGNMENTS: CIRCUIT PROTECTION DEVICES

12-5A Write Full Answers

1. Name the three conditions that can damage or destroy an electric circuit.
2. Define the term overload and explain how it can occur.
3. How can overload damage be prevented?
4. Describe the possible results if a dangerous overload is allowed to exist for some time.
5. What is a short circuit?
6. Describe how an overload can develop into a short circuit and cause a fire.
7. Why is the current rating of a fuse of such great importance in protecting electric circuits?
8. How can buildings and electric circuits be protected against lightning?
9. Name the main part of a fuse and explain its purpose.
10. List four types of fuses and state their typical use and ratings.
11. Describe the internal construction of a plug fuse. Make a labelled drawing to illustrate your answer.

12. Name the three main types of circuit breakers.
13. What is the basic purpose of lightning rods and lightning arresters?
14. Explain how a lightning arrester works.

12-5B **Indicate True or False**

1. An overload can be avoided by increasing the size of the fuse protecting the circuit.
2. An overload is dangerous because it can melt the insulation of the circuit wires and cause a short circuit and fire.
3. The current rating of a fuse is of no importance in the protection of an electric circuit.
4. The fuse link of a plug fuse is made of a fusible alloy .
5. A fuse link broken by an overload current will not be vapourized.
6. Plug fuses are available with current ratings up to 60A and voltage ratings of 125V and 250V.
7. Cartridge fuses are used for major appliance circuits.
8. A fuse or circuit breaker is always connected in series with the circuit or load it protects.
9. A thermally actuated circuit breaker uses an electromagnet as its current-sensing element.
10. Lightning rods and arresters work on the principle of directing the atmospheric electricity into the earth.

12-5C **Select Correct Answer**

1. An overload may be caused
   a) by a 15A fuse
   b) by a fuse that is too large
   c) when the voltage drops below the designed value
   d) if more than one appliance is plugged into a receptacle
2. A prolonged overload current
   a) can easily be avoided with a 30A fuse
   b) will not affect a circuit at all
   c) may melt the insulation, cause a short circuit, and start a fire
   d) usually cools off and corrects itself
3. A short circuit
   a) is a path of extremely low resistance that bypasses the load
   b) cannot occur in a line protected by a fuse
   c) usually melts the fuse link quite slowly
   d) can occur only in twisted linecords
4. Atmospheric electricity
   a) is dangerous only when lightning strikes
   b) cannot be stopped from doing damage
   c) will always produce a lightning bolt
   d) can damage or destroy electric circuits and equipment
5. An electric circuit
   a) usually has more than one fuse to protect it
   b) has a specific voltage rating and a maximum allowable current rating
   c) needs no specific overload protection
   d) is unaffected by heavy load currents
6. A common receptacle
   a) should never be used for more than one appliance
   b) can supply up to three appliances
   c) should always be protected with a 30A fuse
   d) should always be protected with a 20A fuse
7. The main part of a fuse
   a) is its glass case
   b) is its medium screw base
   c) is the contact in the centre of its base
   d) is the fuse link made of fusible alloy
8. Plug fuses are available with
   a) any desired current and voltage rating
   b) current ratings of 10, 15, 20, 25 and 30A

c) voltage ratings of 125, 250 and 500 volts
d) current ratings from 1A to 60A

9. Disposable cartridge fuses are generally used for
   a) all outlet circuits in a home
   b) all heavy-duty outlet circuits
   c) major appliance circuits
   d) minor appliance circuits

10. A lightning rod
    a) can be installed anywhere on a building
    b) discharges atmospheric electricity into the earth
    c) arrests lightnings by short circuiting them
    d) can be made of any material, in any desired shape

13

WIRING DEVICES AND MATERIALS

The professional electrician is a highly skilled tradesman whose time and labour are valuable and must therefore be used in the most effective way. Electrical manufacturers have recognized this fact and have developed wiring devices and materials that speed up the construction of electric circuits. These new materials relieve the electrician of much unnecessary work and result in safer and more reliable installations.

13-1 CSA APPROVAL

To ensure public safety from electrical and fire hazards, all wiring devices and materials must be **approved by the Canadian Standards Association**, or **CSA**.

Figure 13-1 CSA Approval on Electrical Device

The CSA maintains extensive testing laboratories. Each new electrical device is thoroughly tested before it is allowed on the Canadian market. To obtain CSA approval, electrical devices must conform with the provisions of the **Canadian Electrical Code (CEC)** and the various provincial regulations governing electrical equipment.

Figure 13-2 CSA Approval on Package

Most countries have testing laboratories and a national electrical code similar to those in Canada. In the United States, the **Underwriters Laboratories** and the **National Electrical Code** govern the design, production and installation of electrical circuits and equipment.

13-2 SOLDERLESS WIRING CONNECTORS

There are four basic types of solderless wiring connectors: set-screw connectors, twist-on connectors, crimp-on connectors and bolt-on connectors. Each type has specific advantages for particular applications.

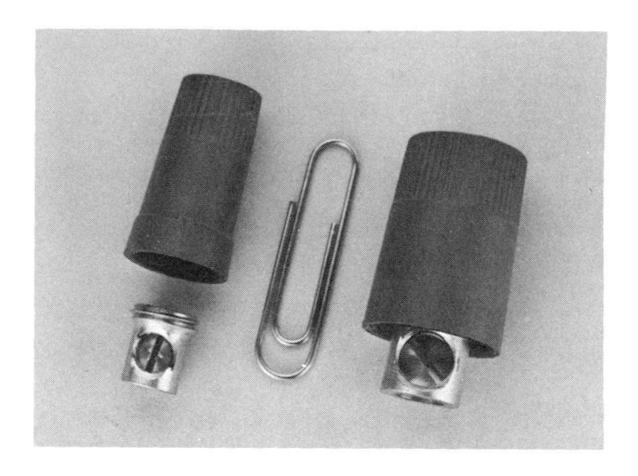

Figure 13-3 Set-Screw Connectors

Set-screw connectors consist of two parts: a brass fitting with a set screw for joining the conductors, and a plastic shell for insulating and protecting the finished splice. Figure 13-4 shows how conductors are joined in a set-screw connector before the insulator shell is mounted on the fitting.

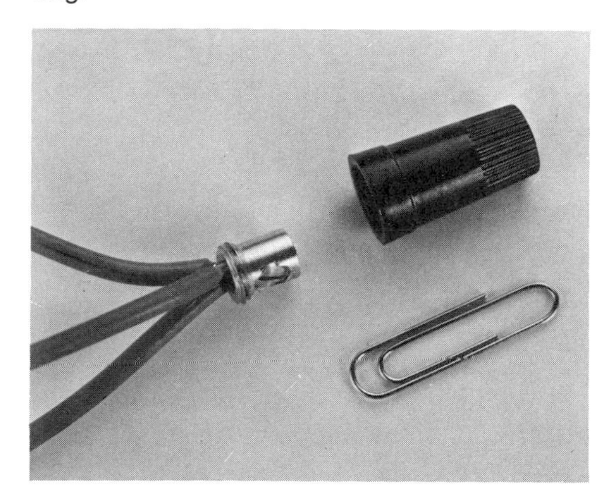

Figure 13-4 Joining Conductors with a Set-Screw Connector

Twist-on connectors consist of an insulator shell containing a conical metal spring. The bare ends of the conductors which are to be joined are pushed into the connector and the insulator shell is twisted clockwise until the wires are firmly joined.

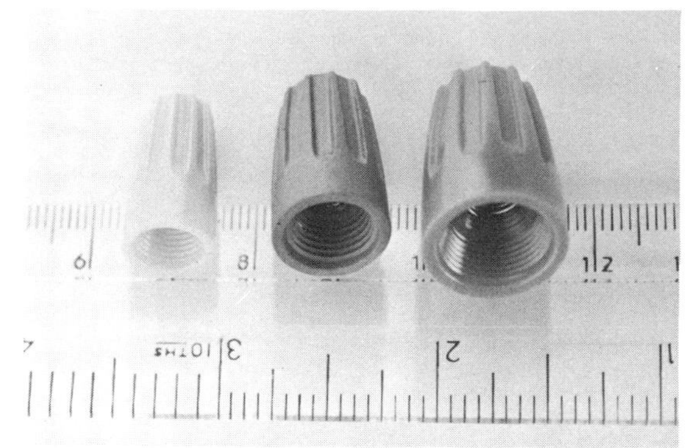

Figure 13-5 Twist-On Connectors

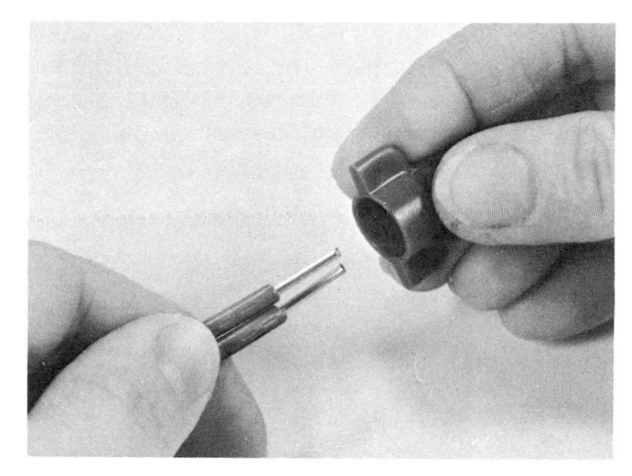

Figure 13-6 Joining Conductors with a Twist-On Connector

The twist-on connectors are the most widely used solderless wire connectors in the home-building industry. They make effective, well-insulated joints in a minimum of time.

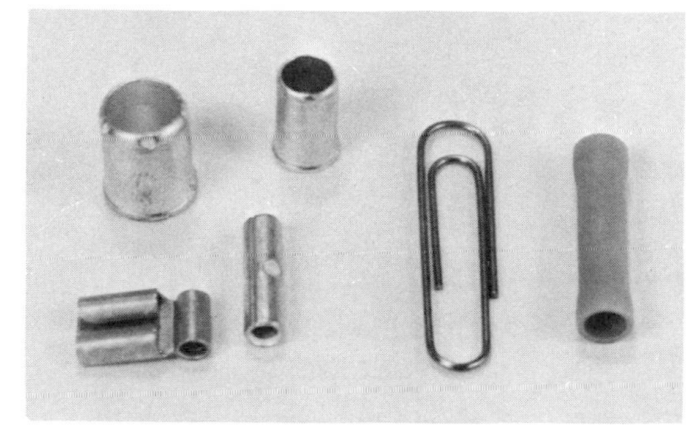

Figure 13-7 Crimp-On Connectors

Figure 13-8 Crimping Tool in Use

Crimp-on connectors require a special **crimping tool** for their installation. These connectors are manufactured in many types and sizes, with and without insulating shells. Their main advantage lies in their small size and in the ease with which they can be installed. Crimp-on terminals for conductors in appliances, automobiles and electrical equipment are especially popular.

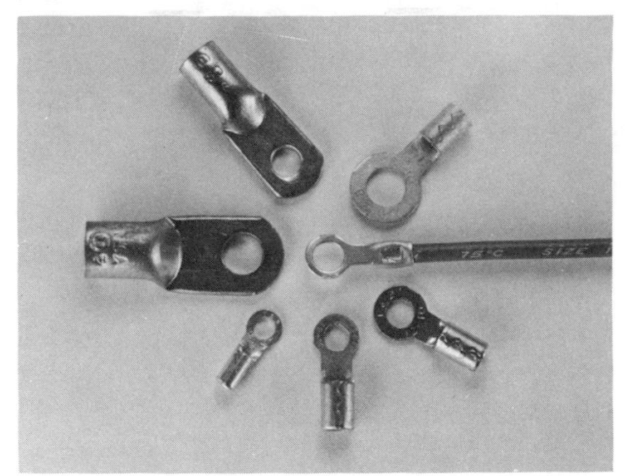

Figure 13-9 Crimp-On Terminals

For joining heavy conductors, **bolt-on connectors** are used. These connectors are made of solid copper to provide maximum conductivity. After installation, they must be carefully insulated with electrical insulating tape.

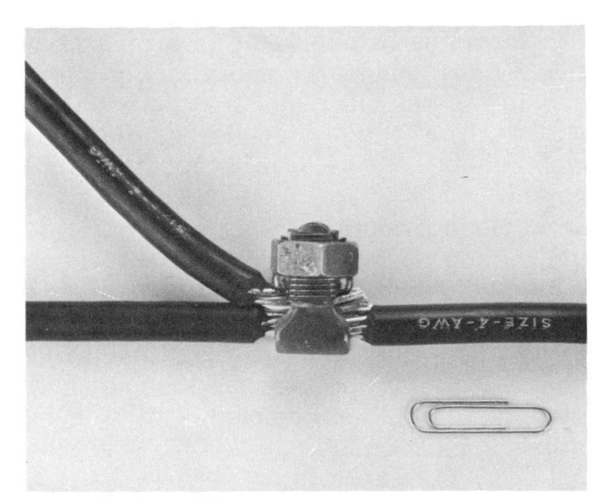

Figure 13-10 Bolt-On Connector Before Insulation

13-3 ELECTRICAL INSULATING TAPES

There are frequent occasions when insulated solderless connectors cannot be used or are not available for making conductor joints. In these cases, hand-formed joints are made that must be insulated with electrical tape.

Vinyl thermoplastic insulating tape is the most widely used electrical tape today. It is moisture-proof and has excellent insulator characteristics.

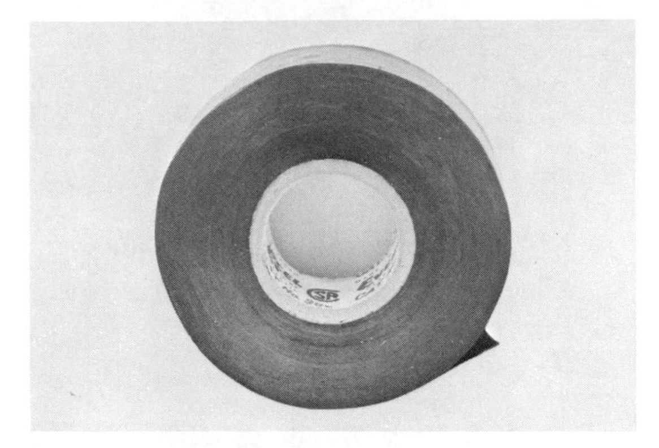

Figure 13-11 Thermoplastic (Vinyl) Insulating Tape

Glass cloth tape (#27) must be used if a conductor joint is exposed to temperatures that would soften or melt thermoplastic tape.

Cotton or **friction tape** is an inexpensive type of insulating tape. Since it is not moisture-proof, it can be used only in certain low-voltage circuits, or for protecting joints insulated with splicing compound.

Splicing compound is used for insulating joints in high-voltage circuits where a considerable thickness of insulation is required. Although splicing compound is an excellent and moisture-proof insulating material, it can be damaged easily and must therefore be protected with a layer of thermoplastic tape.

13-4 LAMP AND APPLIANCE CORDS

Lamps and appliances are connected to a receptacle by means of flexible **linecords.** These cords have two or three copper conductors that are stranded to provide maximum flexibility. The most common wire sizes for linecords are AWG 16 and 18.

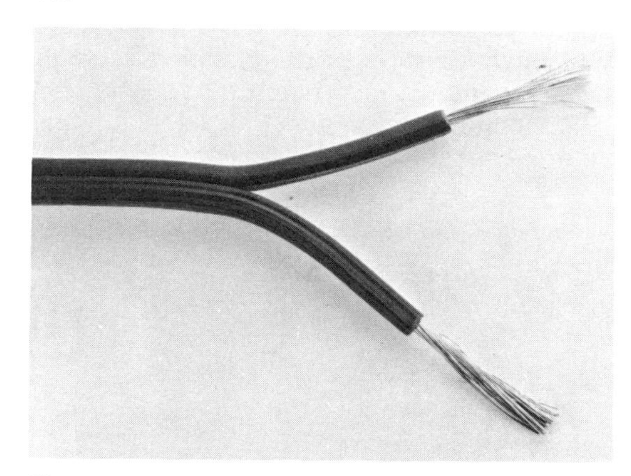

Figure 13-12 Lamp Cord, Type SPT

The simplest type of cord is **lamp cord**, type **SPT** or **SP**. It consists of two parallel conductors surrounded by a tough thermoplastic coat (type SPT) or a moulded rubber coat (type SP). Lamp cord is used for lamps, radios, clocks, electric razors and other non-metallic small appliances.

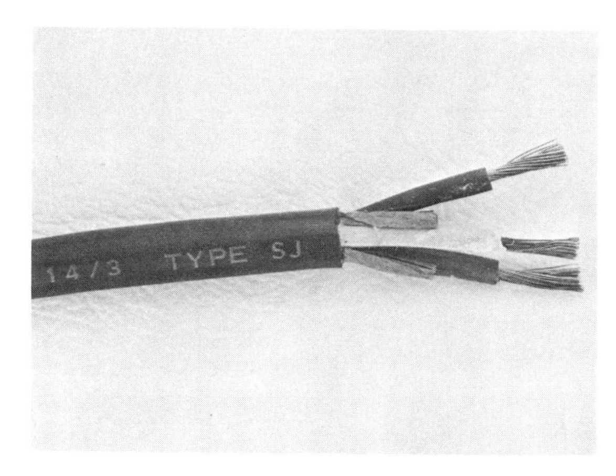

Figure 13-13 Flexible Cord, Type SJ

For hand-operated tools and similar appliances a special rubber or neoprene insulated linecord is used. This cord is tough and durable. Its two- or three-stranded conductors have rubber insulating sleeves separated by strips of tough paper twine. The whole cord has an outer casing of rubber, type **SJ**, or neoprene, type **SJO**. The neoprene-insulated cord is oil-resistant.

Figure 13-14 Heater Cord

Electric heating appliances such as flat-irons and toasters must have special, **heat-resistant** linecords. Their stranded conductors are rubber insulated and covered with a thick layer of fibrous asbestos. The whole cord is protected by a tough outer case of woven cotton.

13-5 PLUG CAPS, APPLIANCE CONNECTORS AND LAMPHOLDERS

Plug caps for linecords are either separately attached or moulded directly on the flexible cord. Two types of plug caps are in common use: the two-prong plug, and the U-ground plug.

Figure 13-15 Typical Two-Prong Plug Caps

The **two-prong plug cap** is used whenever a ground connection is not required. These plug caps are available in many materials and in several sizes and shapes.

Figure 13-16 U-Ground Plug Cap

The **U-ground plug cap** is generally used for hand-operated power tools and for larger appliances requiring a **ground connection** for their metal casing to protect the user from electric shock. The U-ground plug has a third, U-shaped prong to which the grounding wire must be connected. The three prongs of the U-ground plug are colour-coded for proper connection to a three-wire flexible cord, as shown in Figure 13-17.

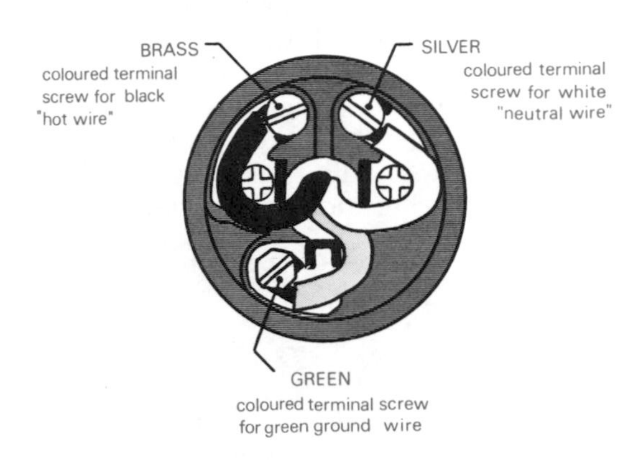

Figure 13-17 Proper Wiring of a U-Ground Plug Cap

Linecords for heat-producing appliances, such as flat-irons and electrical skillets, frequently are equipped with a detachable **appliance connector** made of heat-resistant bakelite. The modern designer tends to save the cost of this connector by wiring the linecord directly to the appliance.

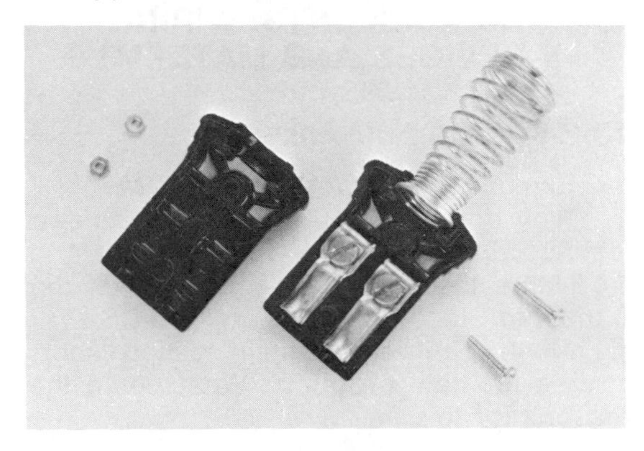

Figure 13-18 Appliance Connector, Disassembled

The **medium-base lampholder** shown in Figure 13-19 is the main part of most electric lamps and lighting fixtures. The holder consists of a porcelain or bakelite medium screw base with two terminal screws, a cap with set screw for attachment to a lighting fixture, a cardboard insulating sleeve, and a brass outer shell. The holder is available with or without a built-in switch.

The lampholder can be snapped apart at the point at which the shell is attached to the cap. When a lamp cord is connected to the terminal screws of the base, care must be taken to cover the terminal screws with the cardboard insulator when the holder is snapped together.

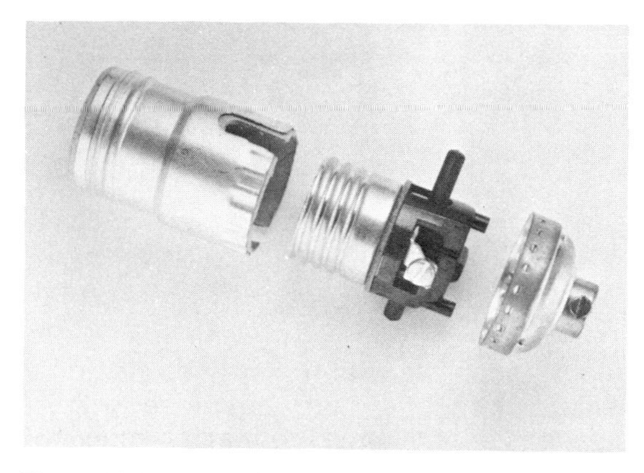

Figure 13-19 Lampholder with Built-In Switch, Disassembled

13-6 ASSIGNMENTS: WIRING DEVICES AND MATERIALS

13-6A Write Full Answers

1. Explain the purpose of the CSA.
2. Why does the CSA maintain extensive testing laboratories?
3. Name the American equivalent of the Canadian Electrical Code and the CSA.
4. Most countries maintain testing laboratories for new products. What could the reasons for this be?
5. Name the four basic types of solderless wiring connector.
6. In an electrical catalogue, check the types and sizes of solderless connectors available to the electrician. Make a chart showing this information.
7. Name the four principal kinds of electrical insulating tape. Explain the physical characteristics and main uses of each one.
8. Why must friction tape never be used in high-voltage circuits?
9. Describe the construction and preferred use of the main types of flexible linecords.
10. Metal-encased hand-operated power tools must be equipped with U-ground receptacles. Give the reason for this CEC regulation.
11. Explain the purpose of the insulating sleeve inside a standard lampholder.

13-6B Indicate True or False

1. All new electrical devices must be approved by the CSA before they can be put on the Canadian market.
2. The CSA is not bound by the provisions of the Canadian Electrical Code.
3. An American equivalent of the CSA is the UL.
4. The chief aim of the CSA testing procedure is the protection of the manufacturer.
5. With set-screw connectors, the bare conductors must be twisted before they are inserted into the connector fitting.
6. Twist-on connectors are the most popular connectors in house wiring.
7. When using twist-on connectors, the electrician must twist the conductors before he inserts them in the device.
8. Crimp-on connectors are larger than twist-on connectors and therefore more expensive to use.
9. Bolt-on connectors can be used for conductors of any size.
10. Friction tape must never be used except in low-voltage circuits.

13-6C **Select Correct Answer**

1. The CSA approval
   a) is not needed for hand-held appliances
   b) warrants electrical devices against mechanical defects
   c) certifies that electrical devices are safe for use by the Canadian public
   d) certifies that electrical devices can be used anywhere in the world
2. An American equivalent of the CSA approval is
   a) the National Electrical Code
   b) issued by the Underwriters Laboratories
   c) acceptable in Canada without further testing
   d) not required in Canada
3. Solderless wiring connectors
   a) must be insulated after installation
   b) are available only for conductors smaller than AWG 10
   c) require special tools for their installation
   d) save time and labour and result in more reliable joints
4. Bolt-on connectors
   a) are made of pure copper to provide maximum conductivity
   b) fit all conductor sizes, from AWG 4/0 to AWG 36
   c) have their own insulator shells
   d) can be used in place of twist-on connectors
5. Electrical insulating tapes
   a) can be used for any conductor joint
   b) are similar in their insulating characteristics
   c) must be carefully selected to match voltage, temperature and humidity conditions
   d) do not need CSA approval
6. Lamp cord, type SPT
   a) can be used with any hand-operated power tool
   b) must be used only for small, non-metallic appliances and for lamps
   c) can be fitted with an appliance connector and used as heater cord
   d) is available with three conductors and a rubber coat
7. Type HPD or HPN linecord
   a) has thermoplastic insulation
   b) can be used only for power tools
   c) must never be used for heat-producing appliances
   d) is heat-resistant and used for heater cords
8. A U-ground plug cap
   a) provides grounding for electrical appliances
   b) has a third prong to give it greater stability
   c) can be connected to any kind of linecord
   d) has differently coloured terminal screws to improve its appearance
9. A medium-base lampholder
   a) cannot be taken apart once assembled
   b) must be carefully checked for proper insulation before assembly
   c) cannot be attached to a fixture
   d) accepts lamps of any type and size

14

SIGNAL DEVICES AND SIGNAL CIRCUITS

Signal circuits are low-voltage, low-current circuits. They have less insulation, thinner conductors, and a less rugged design than standard power and lighting circuits.

The Electrical Safety Code contains special regulations for the wiring of signal circuits. Signal wires must not be placed in the same enclosure as power-circuit conductors unless special provisions are made to isolate the two types. It is therefore common practice to wire signal circuits entirely apart from power circuits to reduce the possibility of fire or electric shock.

14-1 SIGNAL-CIRCUIT COMPONENTS

The components required for a common signal circuit are few and simple: a power source—usually a bell transformer—some signal wire or cable, a push-button switch, and a bell or chime to sound a signal.

Figure 14-1 Simple Signal Circuit

14-1A The Bell Transformer

Bell transformers are available in two basic types: transformers with an output voltage of **6–10 volts a.c.** for simple bell or buzzer circuits, and transformers supplying **12–18 volts a.c.** for door chimes and annunciator circuits.

Figure 14-2 Typical Bell Transformer

Usually a bell transformer is **permanently connected** to the 115-volt line and draws current at all times, even when not in use. To limit the transformer's temperature to a safe level, the Electrical Safety Code requires that these devices have built-in overload and short-circuit protection.

A bell transformer consumes approximately 1 watt of power when idle, and about 5 watts when the circuit is activated. This 5-watt level cannot be exceeded, even if the transformer's secondary coil is accidentally short-circuited.

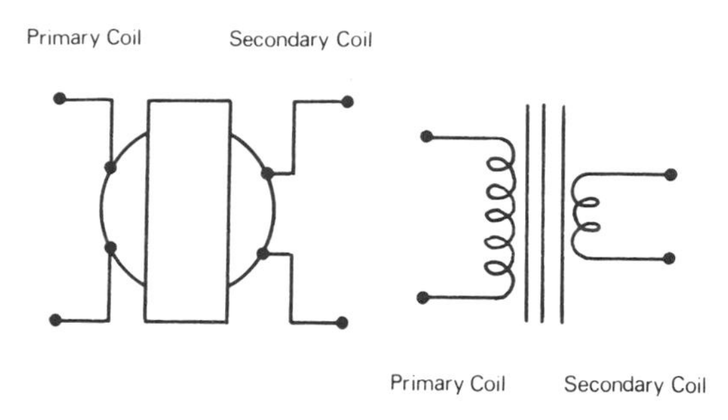

Figure 14-3 Schematic Symbols for Bell Transformer

14-1B Signal Wire and Signal Cable

For new signal circuits, AWG 18 copper wire with thermoplastic insulation is most commonly used. Usually, two or three of these wires are enclosed in a thermoplastic sheath, forming a signal cable.

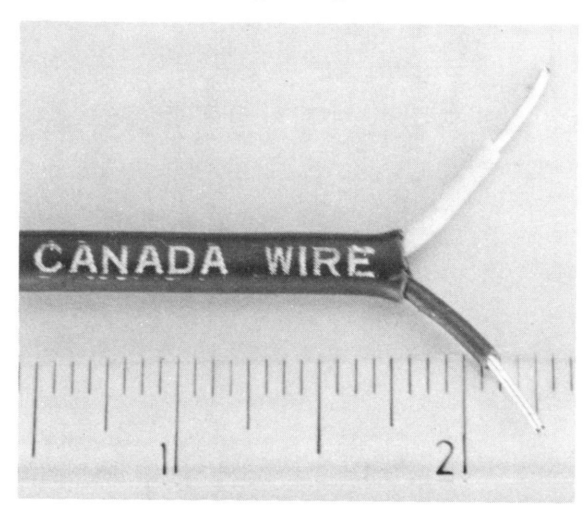

Figure 14-4 Signal Cable

The conductors are **colour-coded** for quick identification of feeder, control and return wire in a signal circuit.

It is customary to staple signal wires or signal cables to their supporting surfaces with the special insulated staples shown in Figure 14-5.

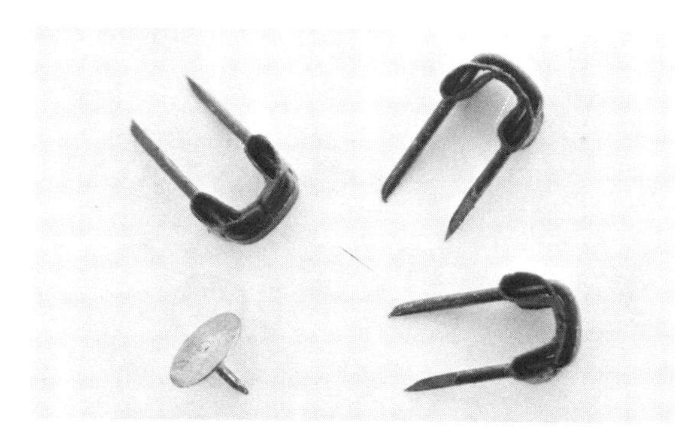

Figure 14-5 Insulated Staples for Signal Wire

14-1C The Push-Button Switch

Signal devices generally are activated by means of spring-loaded push-button switches. These switches have specially designed contacts which keep the circuit closed only as long as the push button is held in the down position. At the instant the button is released, the contacts spring open and break the circuit.

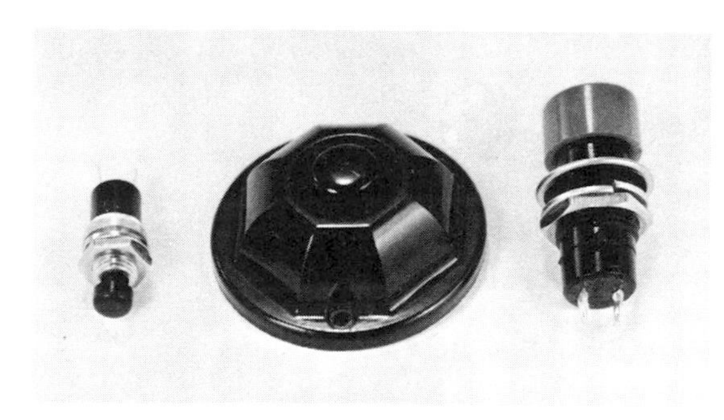

Figure 14-6 Push-Button Switches

Figure 14-7 Internal Construction of a Push-Button Switch

Two terminal screws are provided to connect the signal wire to the switch contacts.

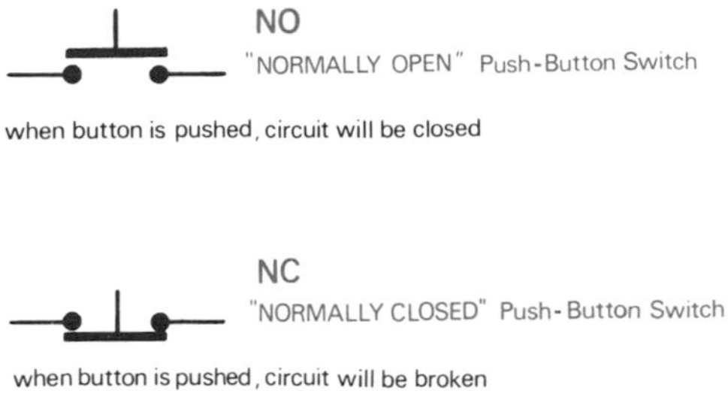

NO

"NORMALLY OPEN" Push-Button Switch

when button is pushed, circuit will be closed

NC

"NORMALLY CLOSED" Push-Button Switch

when button is pushed, circuit will be broken

Figure 14-8 Schematic Symbol for a Push-Button Switch

14-2 SIGNAL DEVICES

The most common signal devices are **bells, buzzers, door chimes** and **annunciators.** Bells and buzzers are preferred for industrial applications where their strident sound can be heard above the usually high noise level. The soft and melodious gong sound of modern door chimes is usually chosen for home signal circuits. Annunciators generally are found in hotels and hospitals where hundreds of signal circuits converge at some central point.

14-2A The Doorbell

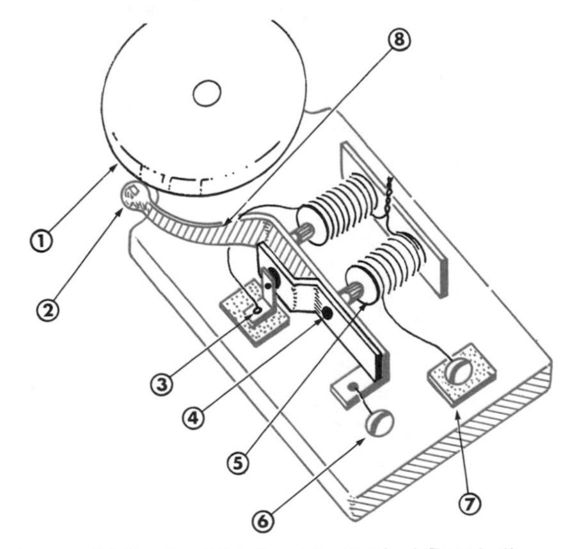

Figure 14-9 Construction of a Typical Doorbell

The doorbell is designed to produce a loud ringing sound when it is activated by an electric current. Its main parts are a U-shaped **electromagnet**, a vibrating iron **armature** with **interrupter contacts**, and a gong.

When the bell is not operating, the steel spring (4) presses its interrupter contact against the stationary contact mounted on the insulated bracket (3). The closed contacts provide a continuous electron path from the uninsulated terminal screw (6) through the two parts of the electromagnet (5) to the insulated terminal screw (7). When an electron current is sent through this circuit, the electromagnets (5) attract the soft-iron armature (8), causing the interrupter contacts to open.

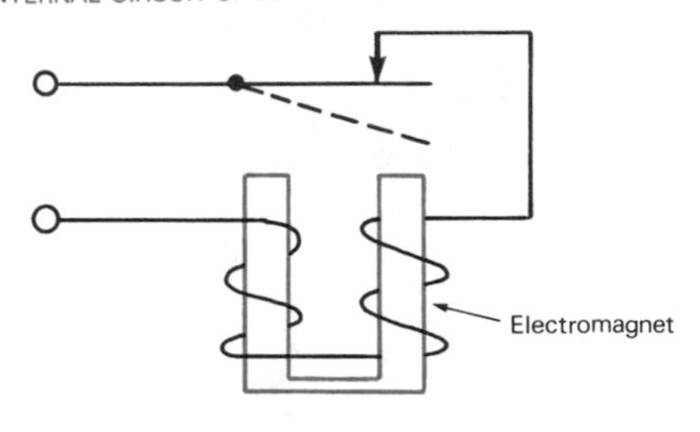

SCHEMATIC SYMBOL FOR BELL

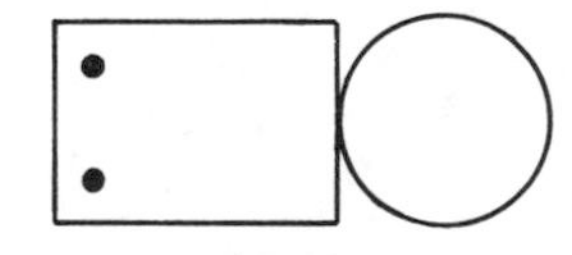

Figure 14-10 Electric Circuit and Schematic Symbol of a Doorbell

At the instant that the circuit is broken the current flow stops, the electromagnets lose their force, and the armature returns to its former position. The interrupter contacts are again closed, and the whole cycle is repeated.

The rapidly repeated sequence of attraction and release of the armature causes strong vibrations. The striker (2), which is part of the armature, hits the gong (1) and produces a loud ringing sound. The intensity of this signal can be adjusted by bending the striker toward or away from the gong.

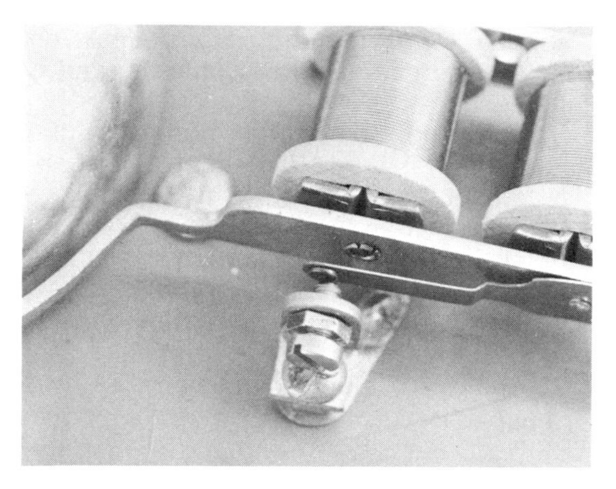

Figure 14-11 Close-Up Photo of Interrupter Contacts

14-2B **The Buzzer**

As its name implies, the sound made by this device is a subdued buzz that is usually confined to one room.

Figure 14-12 Buzzer

The vibrating mechanism and electric circuit of the buzzer are identical to that of the doorbell; the only difference is the absence of the gong and striker. The buzzing sound is produced by the iron armature hitting the poles of the electromagnet. The schematic symbol for a buzzer is shown in Figure 14-13.

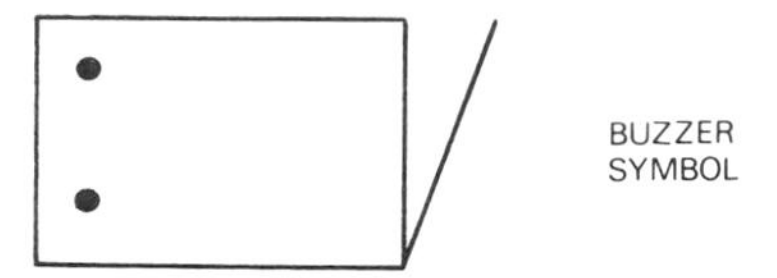

Figure 14-13 Schematic Symbol, Buzzer

14-2C **The Buzzabell**

This is a special signal device combining a bell and a buzzer on the same mounting plate. The buzzabell allows its user to receive and distinguish two different signals: a ring from the front entrance, and a buzz from the backdoor of a residence.

Figure 14-14 Buzzabell, Cover Removed

A buzzabell has three terminal screws: one for the bell, one for the buzzer, and a common return terminal for both; its schematic symbol is shown in Figure 14-15.

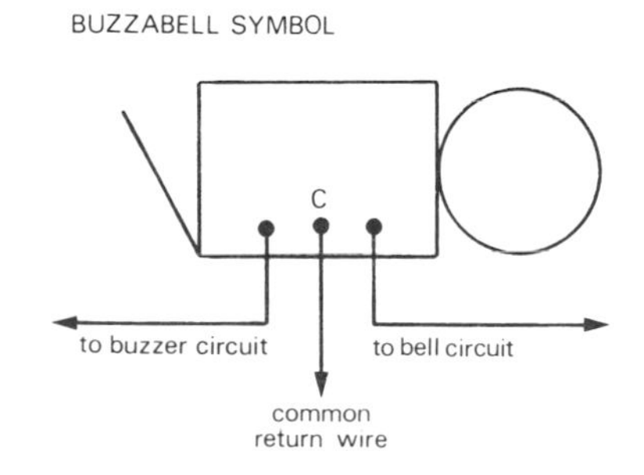

Figure 14-15 Schematic Symbol, Buzzabell

14-2D The Door Chime

Door chimes are made in many sizes and shapes and with a great variety of gong sounds. In most homes, a two-tone chime is installed which produces a gong sound of different pitch for front and backdoor; it replaces the older buzzabell.

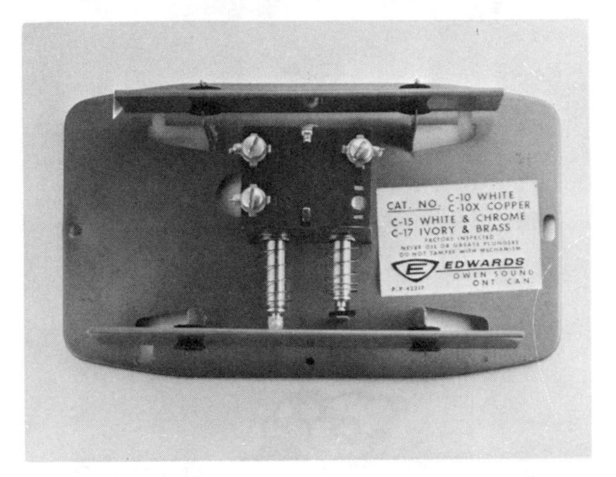

Figure 14-16 Two-Tone Door Chime, Cover Removed

The main part of a door chime is a **solenoid** with a spring-loaded, movable **plunger**. When the solenoid's coil is energized by an electron current, the plunger is attracted into the coil and struck against a tuned metal bar or tube. The impact causes the metal to produce a melodious gong sound. The instant the current flow is interrupted, the spring returns the plunger to its former position.

A typical frontdoor-backdoor chime has two solenoids, each striking a metal bar tuned to a different pitch.

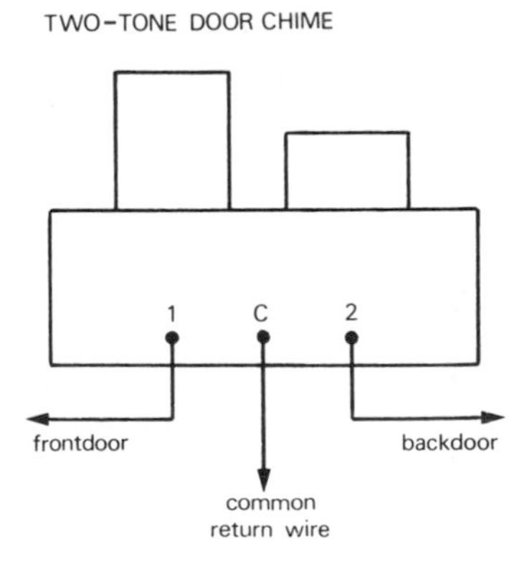

Figure 14-17 Schematic Symbol for Two-Tone Door Chime

14-2E The Annunciator

An annunciator serves two main purposes: it makes a buzzing sound when activated, and it indicates the number of the room from which the signal was sent. It will hold that number until an attendant resets the device.

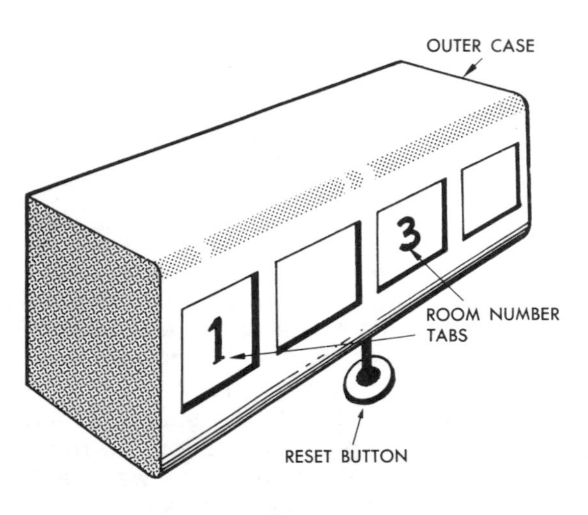

Figure 14-18 Four-Point Annunciator

Annunciators are used in signal systems where many callers must wait their turn for service or assistance. Hospitals, hotels and factories are typical users of annunciator installations. Often dozens, even hundreds of signal lines converge in a single control station, and annunciators line the walls in long rows.

A single annunciator device may serve four, five, ten or more rooms or signal stations. A simple four-point (four-room) device is shown in Figures 14-18 and 14-19. It consists of four electromagnets operating four movable number tabs behind small glass windows. A buzzer is included.

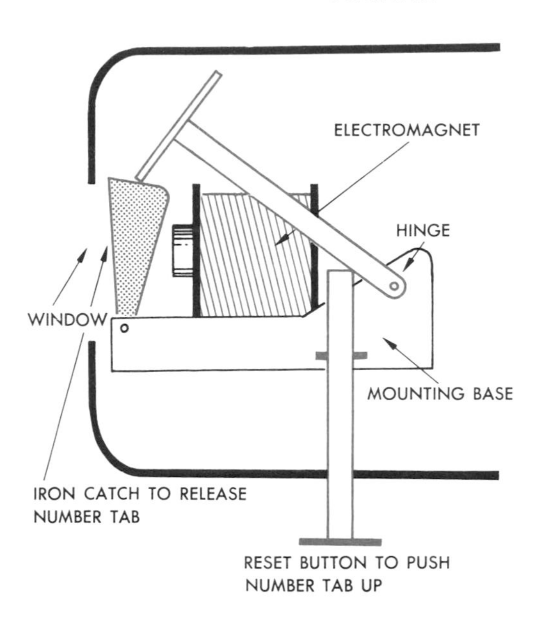

Figure 14-19 Four-Point Annunciator, Cover Removed

At the instant one of the electromagnets is energized by a signal current from a distant signal station, it releases a number tab which falls into the space behind a small glass window, indicating the room number from which the signal was sent. The buzzer sounds at the same time. When the signal has been acted upon, the attendant **resets** the device by pushing its reset button, thereby returning the number tab to its no-signal position above the electromagnet.

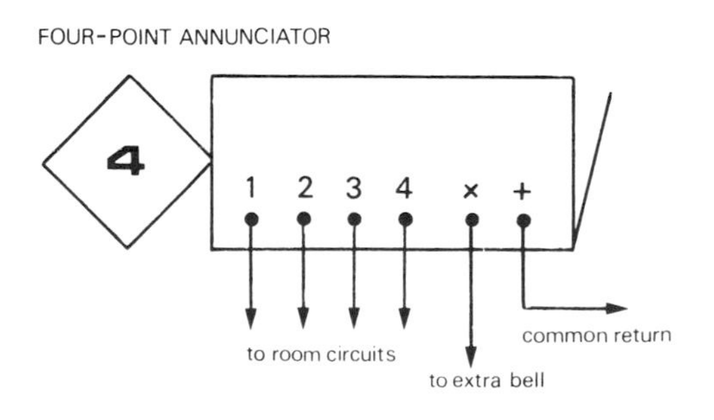

Figure 14-20 Schematic Symbol for Four-Point Annunciator

14-3 SIGNAL CIRCUITS

Except for the connection of the bell transformer primary to the 115-volt line, all parts of a signal circuit must be kept well away from standard power circuits.

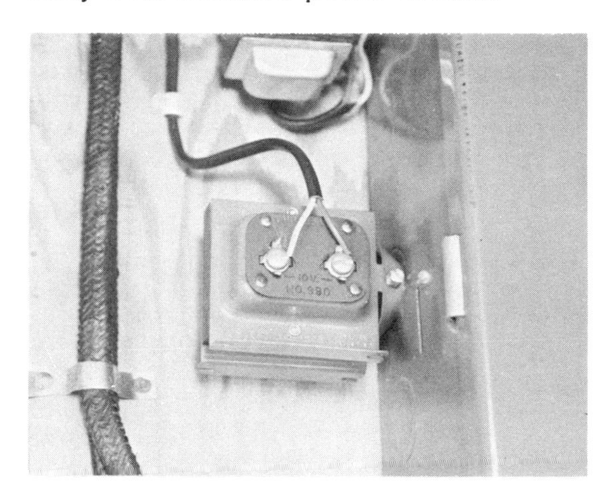

Figure 14-21 Bell Transformer Mounted on Knock-Out Hole of Fuse Panel

The bell transformer usually is mounted directly on a **knock-out hole** of a junction box or on the side of the metal enclosure of the fuse panel. In this way, no part of the 115-volt circuit is exposed, and only the secondary terminals of the bell transformer are accessible (Figure 14-21).

In working with signal circuits (or any other electric circuit), the electrician uses special names for the various wire runs. Thus, the wire running from one transformer terminal to the push-button switch

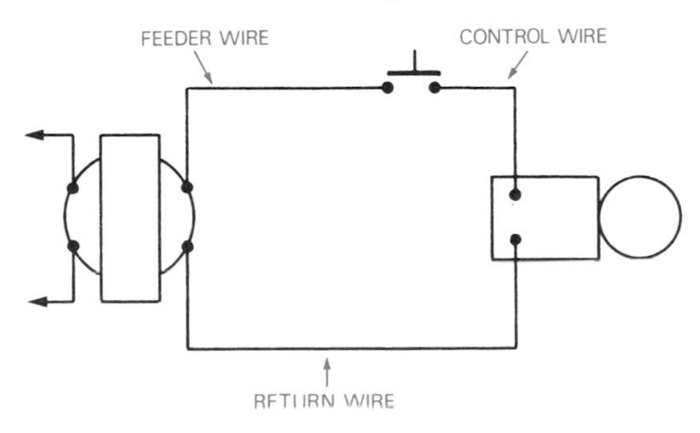

Figure 14-22 Feeder, Control and Return Wires in an Electric Circuit

is called the **feeder wire**; the wire connecting the switch to the signal device, the **control wire**; and the wire completing the circuit from the signal device to the other terminal of the transformer, the **return wire**. In practically all cases, two- or three-conductor signal cable is used.

14-3A A Basic Signal Circuit

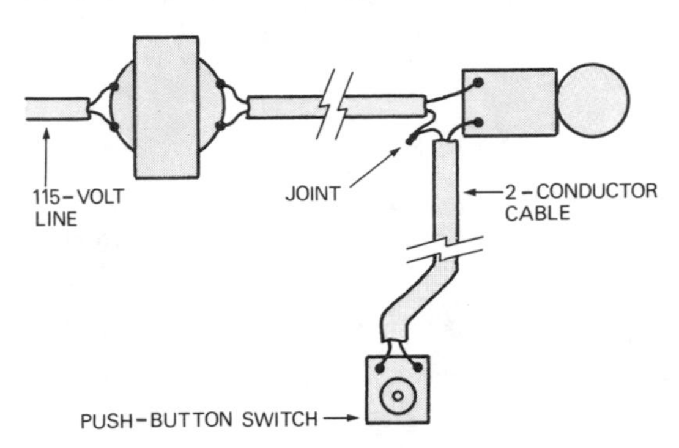

Figure 14-23 Actual Wiring of Basic Signal Circuit

In an actual installation, a two-conductor cable is run from the transformer to the bell, and a second two-conductor cable from the bell to the push-button switch, as shown in Figure 14-23. A twist-on connector joins and insulates the two runs of the return wire.

14-3B Signal Circuit with Controls in Parallel

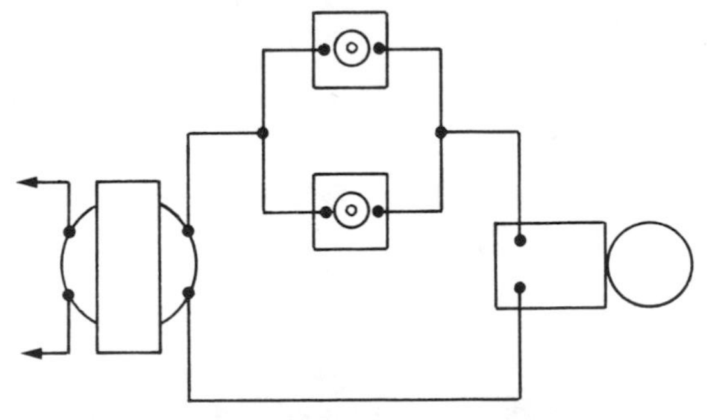

Figure 14-24 Schematic Diagram of Signal Circuit with Controls in Parallel

Controls in parallel may be used where a single bell or buzzer must serve two separate doors, or be activated from two different locations. Each push-button switch can complete the circuit and ring the bell.

14-3C Signal Circuit with Bells in Parallel

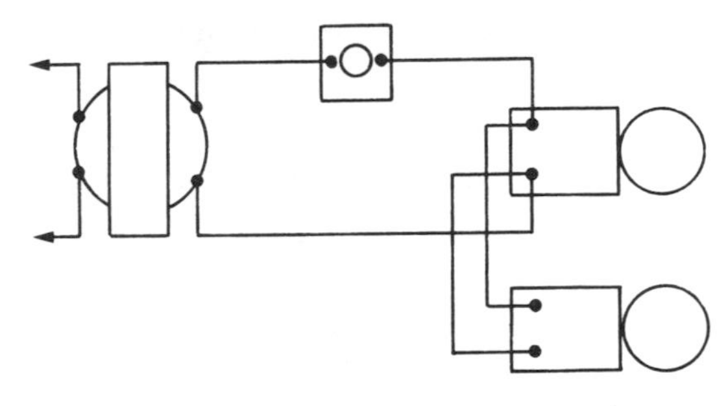

Figure 14-25 Schematic Diagram of Signal Circuit with Parallel Bells

This is a fairly common circuit because in many signal installations a push-button signal must be heard in two or more rooms. As more bells are added to the circuit, however, a more powerful transformer must be installed.

14-3D Signal Circuit with Bells in Series

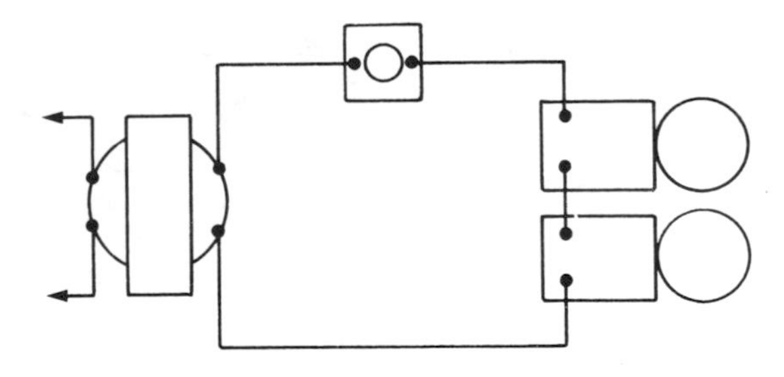

Figure 14-26 Schematic Diagram of Signal Circuit with Bells Connected in Series

There are two disadvantages to this kind of circuit: if either bell is defective, the whole circuit will not work, and, since each bell produces a voltage drop, a transformer with a higher voltage than for a single bell must be used.

14-3E Frontdoor-Backdoor Bell and Buzzer Circuit

This circuit is used when no buzzabell or two-tone chime is available. Each signal device has its own circuit, but both share the transformer and the feeder wire.

14-3F Frontdoor-Backdoor Chime Circuit

This is the most common circuit in home installations requiring push-button switches at two different entrances. A 12–18 volt bell transformer must be installed to supply adequate power for the two-tone chime.

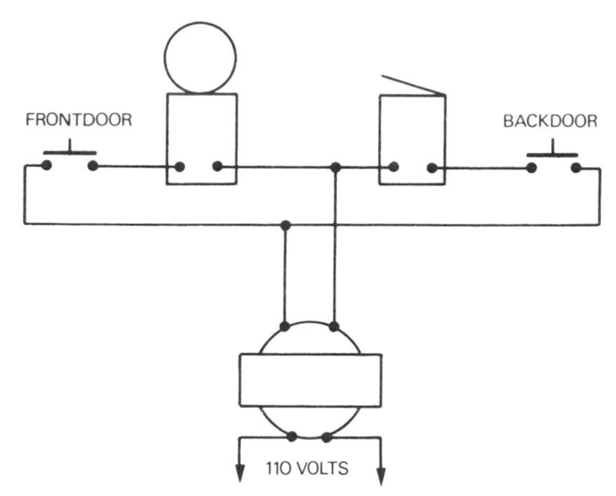

Figure 14-27 Schematic Diagram of Frontdoor-Backdoor Bell and Buzzer Circuit

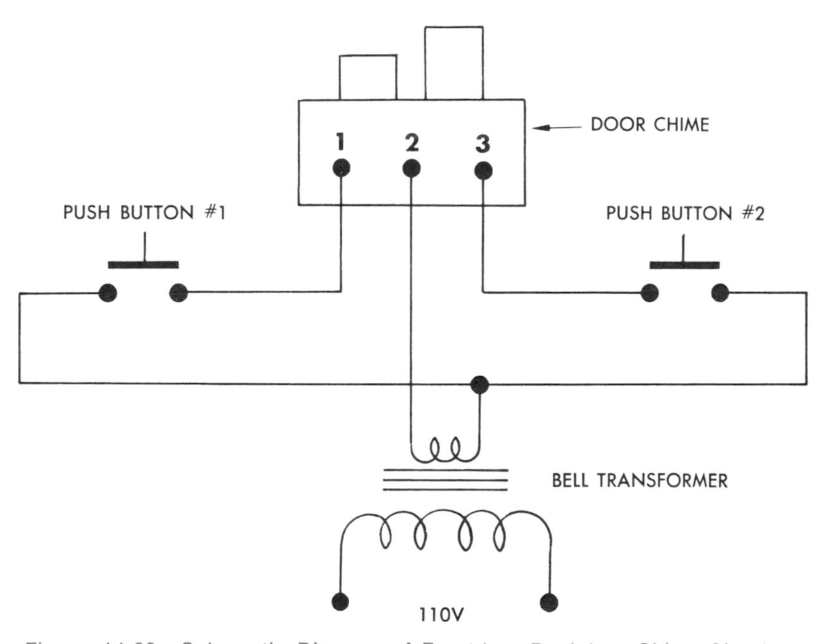

Figure 14-28 Schematic Diagram of Frontdoor-Backdoor Chime Circuit

14-3G Four-Point Annunciator Circuit

This is one of the simplest annunciator circuits, but it has all the main elements of this kind of signal circuit. It differs from a more complex annunciator system in the number of control wires used, and in the size of the transformer.

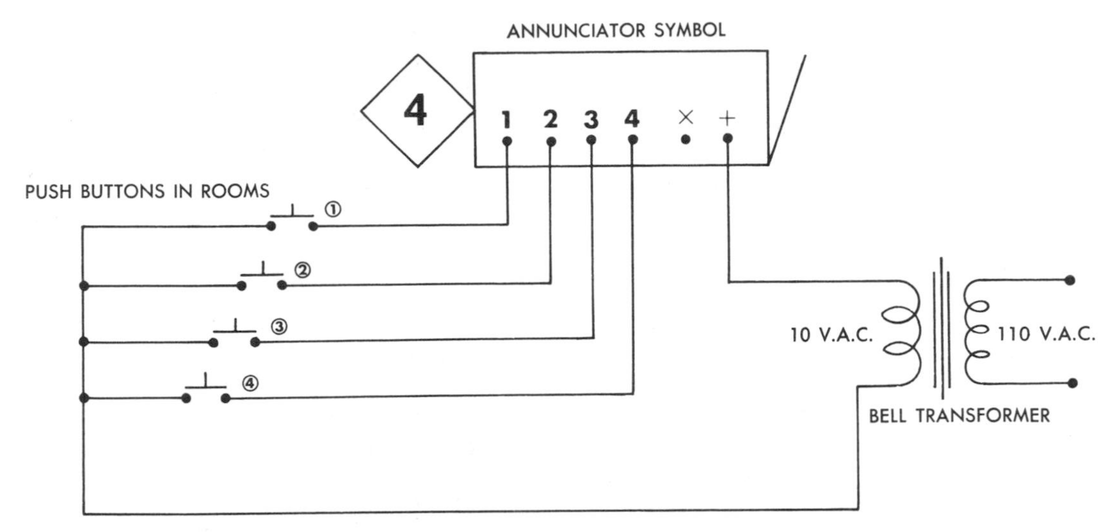

Figure 14-29 Schematic Diagram of Four-Point Annunciator Circuit

14-4 ASSIGNMENTS: SIGNAL DEVICES AND SIGNAL CIRCUITS

14-4A Write Full Answers

1. Why must a signal circuit be kept apart from common power and lighting circuits?
2. List the main components of a simple signal circuit and explain their purpose.
3. Why do signal circuits have thinner conductors and less insulation than standard power circuits?
4. Describe the two basic types of bell transformers and state their main use.
5. The Electrical Safety Code requires that bell transformers have built-in overload and short-circuit protection. What are the reasons for this?
6. How are signal wires and cables attached to their supporting surfaces?
7. Make a drawing (cross-section) showing the internal construction of a push-button switch. Label all parts.
8. Describe in detail how a bell works. Use labelled drawings to illustrate your answer.
9. List the five basic signal devices used in signal circuits. In each case, state the most common use for the particular device.
10. Make a labelled drawing showing the internal construction of a two-tone door chime. Explain how the device works.
11. Describe how a bell transformer is installed and connected to a bell circuit.
12. Draw the schematic diagram for a simple signal circuit. Label all components and wire runs to indicate their purpose.
13. Define the terms feeder wire, control wire and return wire, and state the typical wire colour used for these runs in a signal circuit.
14. Draw the schematic diagram for a frontdoor-backdoor chime circuit and explain how the circuit works.

14-4B Indicate True or False

1. The Electrical Safety Code contains special regulations dealing with signal circuits.

2. Signal circuits may be wired in the same conduits and enclosures as power circuits.
3. All signal circuits operate with 6–8 volts a.c.
4. A bell transformer will not burn up, even when overloaded or short-circuited.
5. A bell transformer consumes approximately five watts of power at all times.
6. Thermoplastic insulation copper wire, AWG 18, is preferred for modern bell circuits.
7. Insulated staples are used to attach signal wires and cables to their supporting surfaces.
8. Annunciators are frequently installed in frontdoor-backdoor signal circuits.
9. The bell transformer usually is mounted directly on a knock-out hole of the fuse panel or junction box.
10. The size and complexity of a signal circuit does not influence the choice of bell transformer used to operate the circuit.

14-4C **Select Correct Answer**

1. The Electrical Safety Code
   a) does not deal with low-voltage, low-current circuits
   b) permits signal circuits in the same enclosures as power circuits
   c) makes no distinction between signal and power circuits
   d) requires that signal circuits be wired apart from power circuits
2. Bell transformers
   a) are available in any desired voltage range
   b) have built-in overload and short-circuit protection
   c) consume about 5 watts of power when idling
   d) consume power only when the signal circuit is activated
3. A push-button switch
   a) breaks the signal circuit
   b) is a spring-loaded, normally closed switch
   c) is a spring-loaded, normally open switch
   d) remains closed once it has been activated
4. The loudness of a bell can be adjusted by
   a) bending the interrupter contacts apart
   b) bending the striker toward or away from the gong
   c) tightening the gong
   d) bending the interrupter contacts together
5. A frontdoor-backdoor chime
   a) uses the same solenoid for both door circuits
   b) has two solenoids and a common sound bar
   c) has a separate sound bar and solenoid for each door circuit
   d) operates best with a 6–8 volt bell transformer
6. The wire runs in an electric circuit
   a) have no special names
   b) cannot be identified individually
   c) need not be distinguished from one another
   d) are called feeder, control and return wires
7. In wiring signal circuits
   a) single wires are preferred
   b) the bell transformer may be installed anywhere
   c) the conductors may not be stapled to their supports
   d) two- or three-wire cables are most often used

PART TWO
APPLIED ELECTRICITY

15

IN THE ELECTRICAL SHOP

The most effective way of gaining a true understanding of electricity is to build, test and analyse live electric circuits. For the beginner, such work can be dangerous, both to himself and to his equipment. The most important rules in the electrical shop is therefore **SAFETY FIRST.**

Accidental short circuits, electric shock and fire are ever-present dangers in working with live circuits. The only real protection against these hazards lies in **your safe work habits.** Your knowledge of shop safety rules and first-aid procedures and your careful attention to these rules are the most important parts of your work in the electrical shop.

15-1 SAFE HABITS SAVE TROUBLE

The safest habit in the electrical shop is to **ask your teacher** for assistance. Make his years of experience with electricity work for you. He can see what might go wrong and recognize dangers that may be hidden to you because of your lack of experience. Use electrical instruments and equipment only with his permission, and ask him to check your experimental circuits **before you turn on the power.**

The electrical shop is not the place for trial-and-error procedures. Electricity moves through an electric circuit with the tremendous speed of **186,000 miles per second.** If the circuit you have built contains an undiscovered fault, you may have no time to save yourself from possible injury, or protect your equipment from serious damage.

So play it safe — always ask your teacher to check your work before you turn on the power.

15-2 ELECTRICAL SAFETY RULES

Electrical Safety in the Home:

1. Replace worn out our damaged line-cords before they cause a short circuit or fire.
2. Keep electrical appliances out of the bathroom. Wet skin conducts electricity painfully well.
3. Avoid touching water pipes of taps and electrical appliances at the same time, especially with both hands.
4. Do not change light bulbs or handle electric appliances while standing on a wet concrete floor.
5. Use only 15-ampere fuses in the fuse panel of your home unless other values are specified.
6. Plug only a single appliance into an outlet to avoid overloading the circuit.
7. Repair electrical equipment only if you are qualified to do so.

Electrical Safety in the Shop:

1. Use electrical instruments and equipment only with your teacher's permission. He is responsible for your activities in the electrical shop.
2. Follow experimental instructions and wiring diagrams as closely as possible.
3. Always compare a completed circuit with the circuit diagram to discover possible wiring faults.
4. Ask your teacher to check a completed circuit before you turn on the power.
5 Never touch a charged wire with your bare hands.
6. Treat every electric circuit as a live circuit until proved otherwise.
7. Handle electric instruments and equipment with care and keep them in good working order.
8. Keep your place of work clean. Remove tools and cut-off pieces of wire from an experimental circuit before you turn on the electricity.
9. Return defective instruments and equipment to your teacher, not to the storage cabinet.
10. Be alert and observant when you work with live circuits. Keep your mind on your work.

Electrical Safety Outdoors:

1. Stay away from fallen power lines! High-voltage electricity can jump at you over a distance of several feet, right through the air!
2. Never climb high-voltage towers.
3. For outdoor work, use only electrical tools that have a grounded outer case, or are specially insulated.
4. Avoid standing under trees during thunderstorms. Seek refuge in a building or in a car.
5. Never be the highest object on a large flat surface during a severe thunderstorm. Electric discharges occur most frequently at high points.
6. Lightning can strike the same place many times, contrary to popular belief.

15-3 DANGER: ELECTRIC SHOCK

Electricity is dangerous and often fatal to all living organisms. The extent of the damage it can do depends on two factors: the **amount of current** flowing through the body, and the **path** of the current.

The amount of electric current which will give a shock to a person varies considerably, but is always quite small. A mere 1/1,000 of an ampere will cause discomfort, and 1/100 of an ampere can kill if it flows through vital organs.

The higher the shock voltage, the more dangerous will be the current level flowing through your body. Any voltage above 40 volts can be a potential killer, and the 115-volt alternating current from a standard outlet can easily give you a severe, even fatal, electric shock.

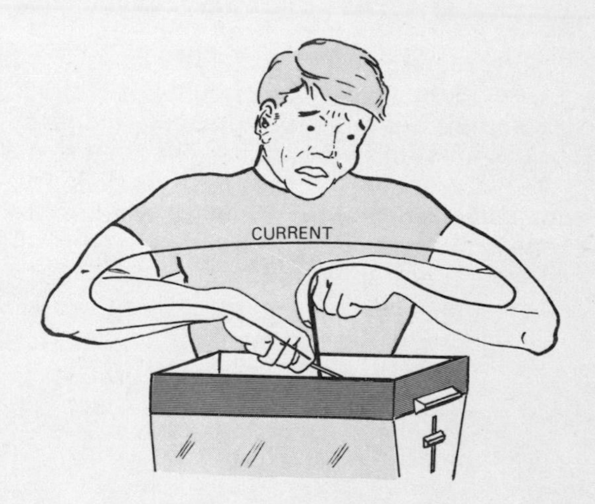

Figure 15-1 A Dangerous Current Path

Even a mild electric shock caused by a current below 0.005 amperes can be quite painful. It will contract the affected muscles so violently that you can be badly hurt if your hand is flung against a sharp edge or a pointed object. A severe electric shock may render its victim unconscious and paralyze his breathing apparatus. If that happens, artificial resuscitation must be started immediately to save the person's life.

15-4 FIRST AID IN CASE OF ELECTRIC SHOCK

1. Do not touch the victim if he is still connected to the conductor that gave him the shock — you may get shocked yourself!
2. **Remove the electricity** from the stricken person. Unplug linecords, remove fuses, throw main switches to the "off" position.
3. Examine the victim immediately after the electricity has been removed. If he is breathing regularly and his heart is beating normally, he likely will recover without further trouble.
4. If he has stopped breathing, **begin artificial resuscitation at once,** using the mouth-to-mouth technique described below.
5. Ask another person to **call a doctor** while you are administering mouth-to-mouth breathing to the victim. If no one else is around, continue artificial resuscitation until somebody shows up.
6. Continue artificial resuscitation until the victim recovers or until the doctor arrives.

15-5 ARTIFICIAL RESUSCITATION

A person who has received an electric shock and has stopped breathing will die from asphyxiation, exactly as in drowning, unless artificial resuscitation is started immediately. Artificial breathing keeps the vital functions of the body alive until the victim recovers from the shock; this may take several hours in severe cases.

The simplest method of artificial breathing is the **mouth-to-mouth technique:**

1. Put the victim flat on his back. Remove any objects which might be underneath him.
2. Tilt his head slightly backwards to straighten out his windpipe.
3. Open his mouth by pulling down his jaw with your left hand. Remove any gum or other object that may hinder the flow of air.
4. Close his nose with thumb and forefinger of your right hand.

Figure 15-2 Examining the Victim of an Electric Shock

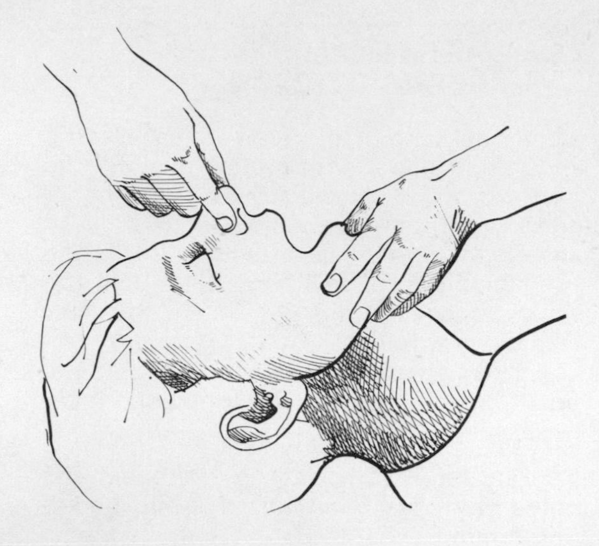

Figure 15-3 Opening the Mouth of a Shock Victim

5. Slowly and carefully blow air into his lungs and watch his chest rise as it fills with air.

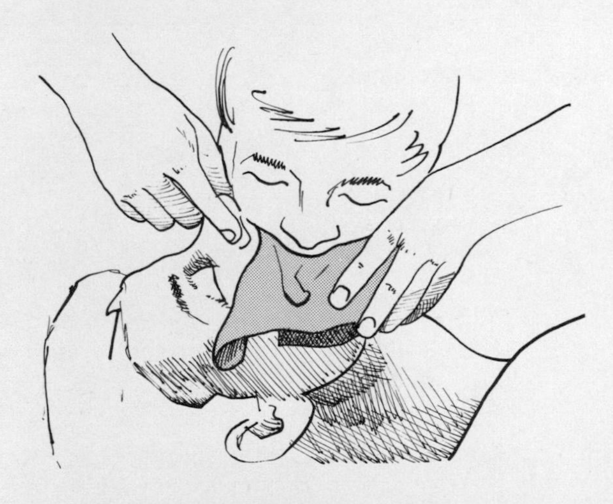

Figure 15-4 Blowing Air Into the Mouth of a Victim of Electric Shock

6. Let the air escape through the victim's mouth and watch his chest fall.
7. Continue blowing air into his mouth at half your normal breathing rate until he recovers.

15-6 HOW TO EXTINGUISH AN ELECTRICAL FIRE

A severe short circuit can set conductors on fire and ignite other combustible materials in its immediate surroundings. If the circuit breaker fails to shut off the flow of electricity, the continuing arc will keep the centre of the fire burning with great intensity.

Figure 15-5 Fire Extinguisher for Electrical Fires (CO_2 Type)

Here is what you should do in case of an electrical fire:

1. Trigger the nearest fire alarm to warn others of the emergency and to alert the fire department.
2. Try to shut off the flow of electricity by unplugging linecords, throwing switches to the "off" position, or tripping the main circuit breakers.
3. Use a carbon-dioxide or dry-powder fire extinguisher to put out the flames. Aim the stream of carbon dioxide at the centre of the fire.

4. **Never use water on an electrical fire.** The stream of water may conduct electricity through your body and give you a severe electric shock. Also, the water will provide new pathways for the electric energy which may cause additional short circuits.

15-7 ASSIGNMENTS: IN THE ELECTRICAL SHOP

15-7A Write Full Answers

1. Why must worn or damaged linecords be replaced as soon as their condition is noticed?
2. What could happen if an electrical appliance is used in a bathroom?
3. Describe what could happen to a person who touches a defective electric kettle with one hand and a water tap with the other hand.
4. Electric water kettles have very short linecords. What could be the reason for this?
5. Why is it dangerous to change a light bulb while standing on a wet concrete floor?
6. Using a fuse with too high a current rating is a dangerous practice. Why?
7. Explain the possible results of plugging several appliances into a single outlet.
8. Why should you ask your teacher's permission before you use electrical equipment in the shop?
9. Name several advantages of following experimental instructions and wiring diagrams as closely as possible.
10. Why should you always compare a completed circuit with its wiring diagram before you allow electricity to flow through it?
11. State several reasons for asking the teacher to check your work before you turn on the power.
12. Explain why touching a charged wire with your bare hands may lead to serious trouble.
13. Why is it necessary to treat every electric circuit as a live circuit until proved otherwise?
14. What damage could a snipped off piece of wire do if it is accidenally left in an experimental circuit?
15. Why must defective instruments and equipment be returned to your teacher, not to the storage cabinet?
16. Describe what could happen if you approach power lines that have been knocked down by a storm.
17. What could happen to you if you use a defective electric drill with a metal casing on wet or damp ground?
18. Why is it dangerous to be the highest point on a large flat surface during a thunderstorm?
19. How can a severe electric shock affect the human body?
20. Describe in detail the first-aid procedure to be used if a person has suffered a severe electric shock.
21. How would you deal with an electrical fire?
22. Water must never be used in dousing an electrical fire. State several reasons for this rule.

16

BASIC ELECTRICAL SKILLS

The construction of reliable electric circuits requires skilful hands. It may seem effortless to you to see an expert joining conductors or soldering a splice, but a great deal of practice is required in learning these skills.

Once you have mastered the basic techniques of working with electricity, the new skills will serve you well for years to come, and in every walk of life.

16-1 MAKING SIMPLE CONDUCTOR JOINTS

Although the new solderless connectors have reduced to a minimum the need for hand-formed conductor joints, the making of a reliable joint still remains one of the most essential handskills in the electrical shop.

The three joints most commonly used for splicing conductors are the rat-tail joint, the tap joint, and the western union joint.

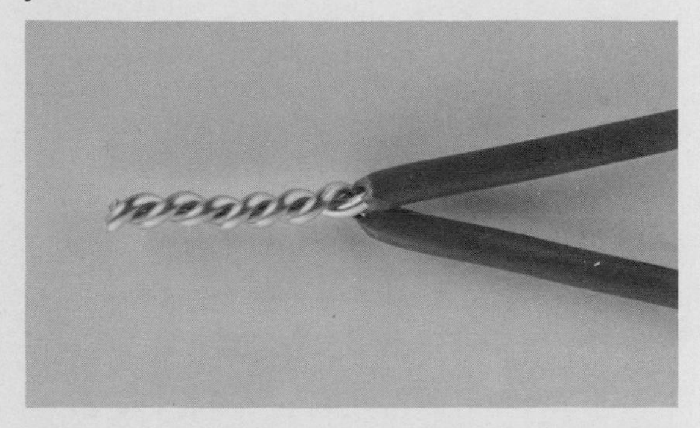

Figure 16-1 Rat-Tail Joint in AWG 14 Conductors

A rat-tail joint is made when two or more conductors must be joined inside a fixture or junction box.

Figure 16-2 Tap Joint

The tap joint is used for connecting a branch conductor to one that "runs through" a box, without cutting the latter apart.

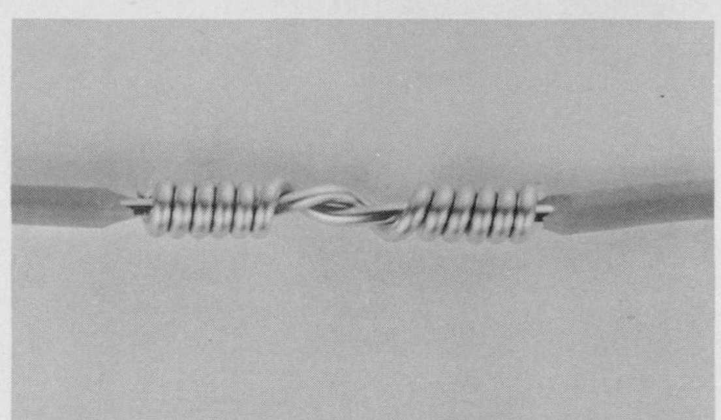

Figure 16-3 Western Union Joint

The western union joint is the strongest of the three types; it is used for splicing breaks in long-run conductors and for extending existing conductor runs.

The Electrical Safety Code requires that conductor joints be "mechanically and electrically secure" and "covered with insulation approved for the purpose."

Mechanically secure means that the conductors must be tightly twisted together, with no loose turns or gaps between the turns. It usually also means that hand-formed joints must be soldered for maximum strength and protection against oxidization.

Electrically secure means that the completed joint must conduct an electric current as effectively as the unbroken conductors. To make sure that this is the case, the conductor surfaces must be absolutely clean, and enough twists or turns must be made to provide sufficient contact area for the electric current to flow from one conductor to the other.

Covered with approved insulation means that the completed joint must be insulated with approved electrical insulating tape. This insulation must be of the same quality and thickness as the original insulation on the conductors.

16-1A Preparing Conductors for Joining

Copper wire is both expensive and in short supply. Try to use as little wire as possible and make sure that all scrap pieces are collected for recycling.

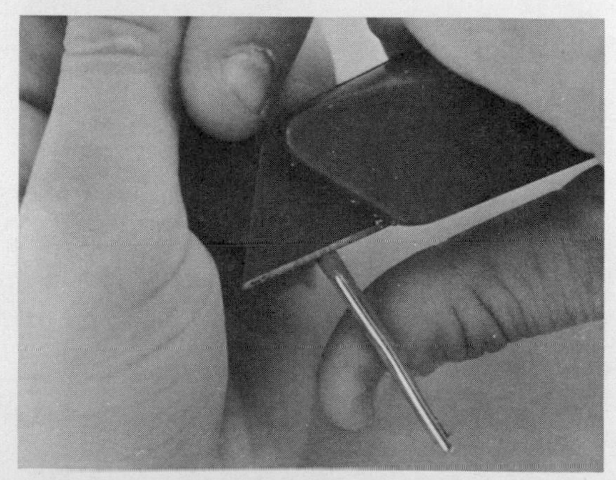

Figure 16-4 Removing Insulation from AWG 14 Conductor

Removing the insulation. The length of insulation to be removed from a conductor depends on the type of joint which you have to make. [Roughly 1½ inches (3.7 cm.) for rat-tail joints, 3 inches (7.5 cm.) for the branch conductor in a tap joint, and about 4 inches (10 cm.) for western union joints.]

Hold the knife at an angle, as shown in Figure 16-4, so that its blade will not cut or nick the soft copper wire, but will glide smoothly along its surface. Cut away one strip of insulation. Then, rotate the wire in your hand and cut away another strip. Continue in this fashion until all the insulation has been skinned off the end of the conductor.

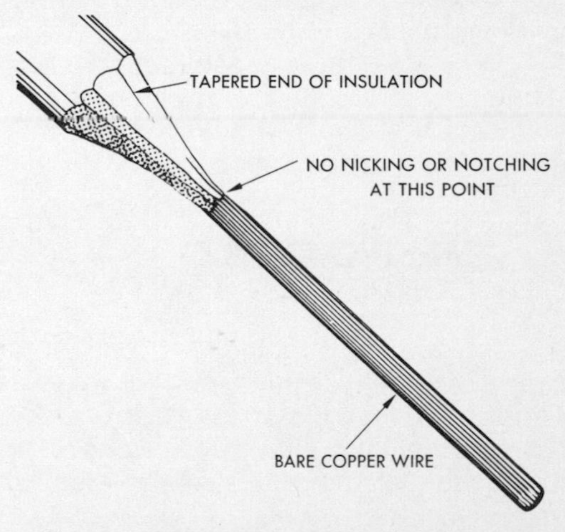

Figure 16-5 Properly Skinned Conductor

Removing kinks and bends. A kink is a sharp bend in a wire. It occurs when coiled-up wire is pulled off a spool without proper care in unwinding the coils. If a kink has been pulled too tight, the wire is damaged beyond repair. A loose kink usually can be straightened out by carefully unwinding it.

Burnishing dull wire is necessary only when old wire is used for making joints. Place the bared, dull end of the conductor on your thumb and scrape its surface clean with the **back edge** of the knife blade. Again be careful not to nick or damage the copper surface.

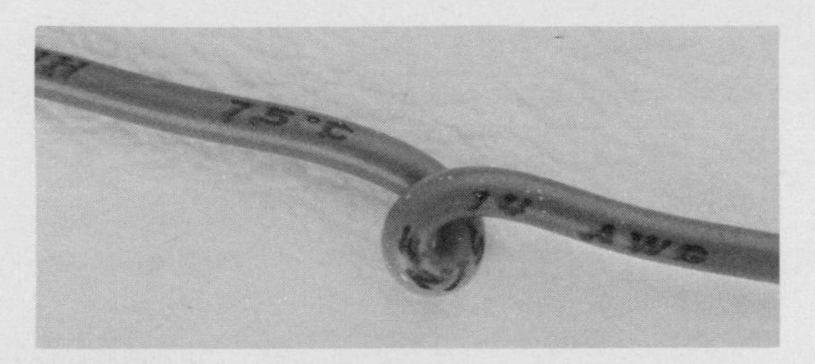

Figure 16-6 A Kink—Damaged Beyond Repair

Assignments:

1. Describe in detail how a conductor is prepared for joining.
2. List the dangers involved in these operations as far as the copper wire is concerned.
3. State the Electrical Safety Code requirements for making conductor joints.
4. Explain how dull or oxidized wire is burnished, and why this operation is necessary.
5. Prepare several lengths of AWG 14 copper wire for joining. Ask your teacher for further instructions. (Prepare one set of wires for a rat-tail joint, a second set for a tap joint, and a third set for a western union joint.)

16-1B **Making a Rat-Tail Joint**

Use the materials listed below to make several rat-tail joints. Ask your teacher to inspect each joint before you make the next one.

Materials:

Two, one-foot (30 cm.) lengths of solid, insulated copper wire, size AWG 14 or 16, for each joint.

Procedure:

Remove 1½ inches (3.7 cm.) of insulation from each wire. Hold the wires tightly parallel with your left hand and bend one end up and the other down, so that the wires will cross ⅛ of an inch (.3 cm.) from the end of the insulation (Fig. 16-7). Now, keeping a firm hold on the wires, twist the ends in a **clockwise** direction with your right hand, forming 6 tight turns. Cut off the wire ends with diagonal cutters.

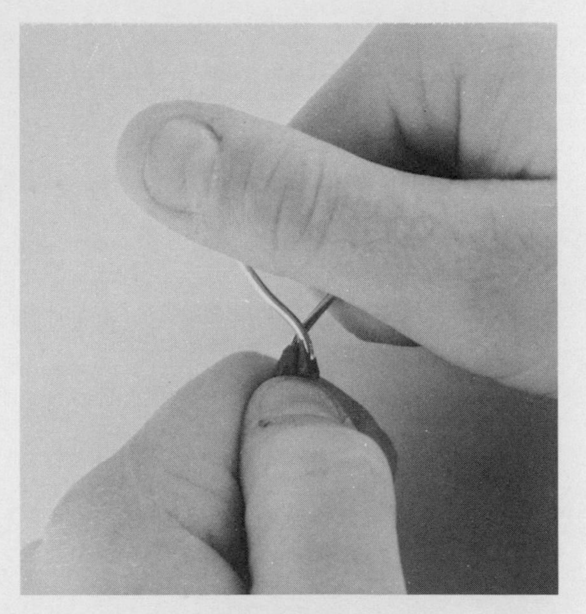

Figure 16-7 Forming a Rat-Tail Joint

The completed joint should be so tightly wound that the wires cannot be moved within it.

If wire sizes below AWG 14 are used, the joint can no longer be formed by hand. Pliers must be used for twisting the wires, but extreme care must be taken not to damage the conductors.

Assignments:

1. Name several typical uses for the rat-tail joint.
2. List the main features of a rat-tail joint.
3. Why should there be at least six twists in a well-made rat-tail joint?
4. The rat-tail joint must not be used in suspended long-run conductors. What could be the reasons for this rule?

16-1C **Making a Tap Joint**

Make several tap joints and ask your teacher to inspect each joint before you make the next one.

Materials:

Two, one-foot lengths of solid, insulated copper wire, size AWG 14 or 16, for each joint.

Procedure:
Remove 1 inch (2.5 cm.) of insulation from the centre of one conductor, and 3½ inches (8.7 cm.) from one end of the second conductor. Hold the bare end of the second conductor tightly against the left side of the gap in the long-run conductor. Bend the bare wire upward, as shown in Figure 16-8.

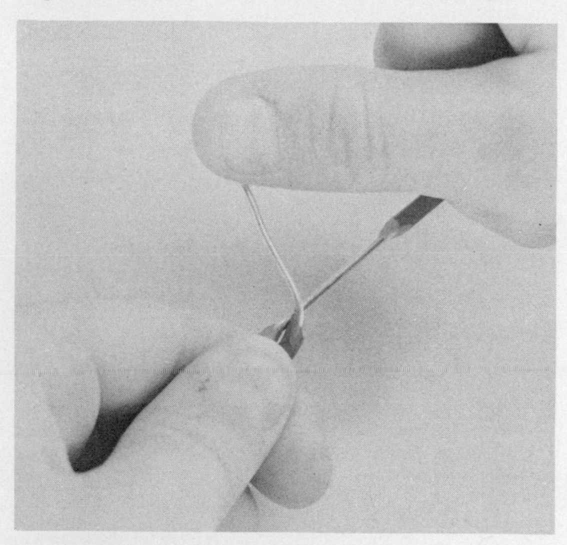

Figure 16-8 Beginning a Tap Joint

Now, without releasing your left-hand hold, make 6 tight turns around the long-wire run. With diagonal pliers, cut off the stubby end of the branch conductor and bend it down carefully.

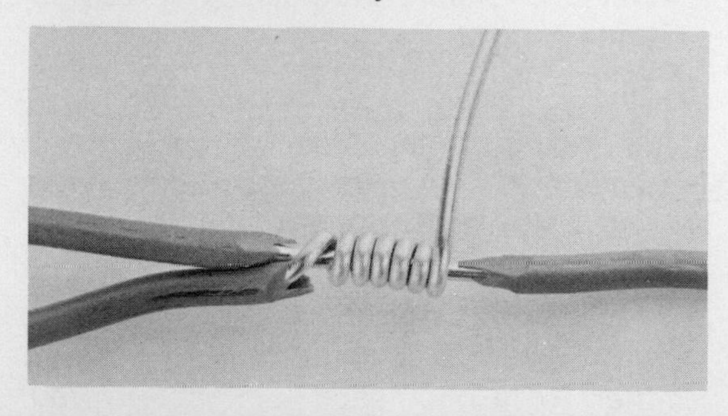

Figure 16-9 Half-completed Tap Joint

Finally, bend the branch conductor downward until it makes a 90° angle with the long-run wire, as shown in Figure 16-2.

Assignments:
1. Make a neat, labelled drawing of the tap joint. Identify the long-run wire and the branch conductor.
2. List the main features of the tap joint.
3. State two reasons for making six tight turns around the long-run wire.
4. Why must you be especially careful in removing the insulation from the long-run wire?

16-1D **Making a Western Union Joint**

This joint is the strongest of all hand-formed joints. It was used originally by the linemen of the Western Union telegraph company to fix breaks in telegraph wires.

Make several western union joints and keep them for soldering later on. Ask your teacher to inspect each joint before you make a new one.

Materials:
Two 18-inch (45 cm.) lengths of solid, insulated copper wire, AWG 14 or 16, for each joint.

Procedure:
Remove 4 inches (10 cm.) of insulation from one end of both wires. Cross the bare ends ¾ of an inch (1.8 cm.) away from the insulation and press them together **tightly** with your left hand (Figure 16-10).

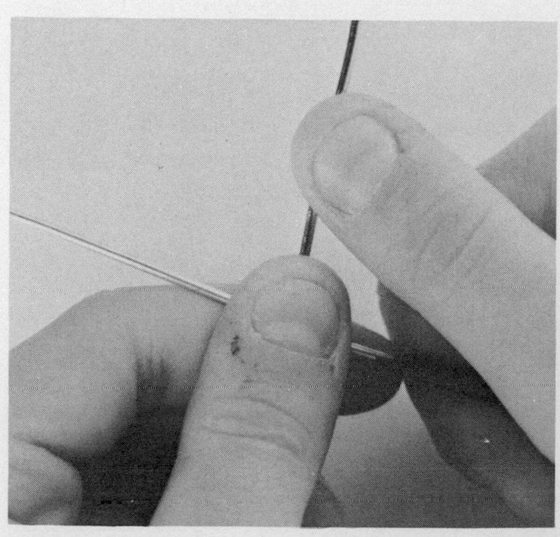

Figure 16-10 Beginning a Western Union Joint

Without releasing your tight hold on the cross-over point, wrap the free end of the left wire tightly around the right wire in a clockwise direction, forming 6 turns. There must be no gaps between the individual turns.

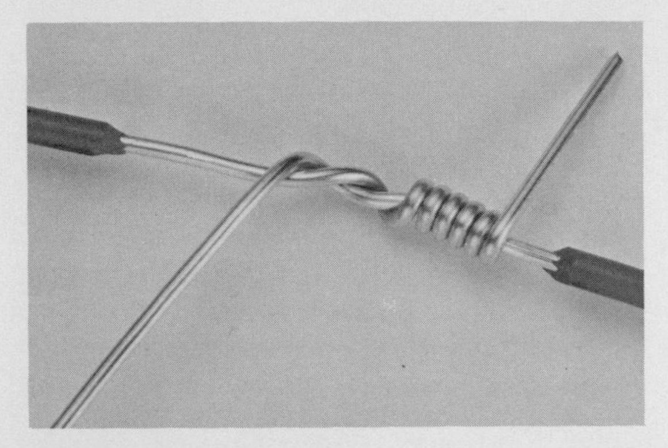

Figure 16-11 Half-completed Western Union Joint

Now, holding the finished turns with your right hand, wind the right wire counterclockwise into 6 tight turns around the left wire.

Cut off the remainder of the wire ends with a diagonal cutter and squeeze them gently into place with a pair of pliers. But be careful not to damage the copper surfaces. The finished joint should be so tightly wound that the wires cannot be moved within the joint.

Assignments:

1. Make a neat drawing of a western union joint. Shade one conductor to distinguish it from the other.
2. List some mechanical and electrical features of the joint and give its principal uses.
3. Why must the wires be tight and unmovable in the completed joint?
4. The western union joint is particularly suitable for mending broken telegraph and telephone lines. Explain the reasons for this.

16-2 SOLDERING

Soldering a joint has three important results: the solder coating prevents the oxygen of the air from reaching the joint, improves the joint's mechanical strength, and increases the contact area between the conductors.

If oxygen is allowed access to a joint, it will combine with the copper to form black copper oxide, making the joint useless. If the joint is soldered, oxygen cannot reach the contact areas, and no oxidation will take place.

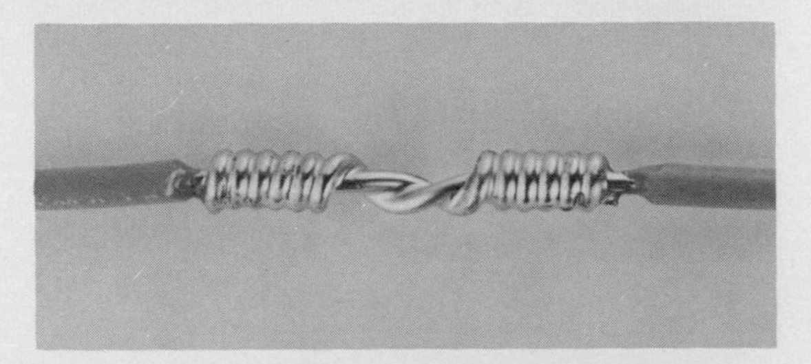

Figure 16-12 Soldered Western Union Joint

The tight bond formed between the copper and the solder metal improves the mechanical strength of the joint.

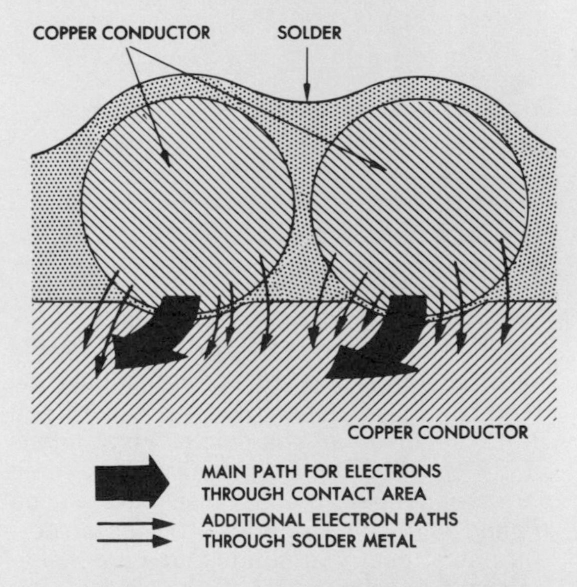

Figure 16-13 Contact Areas in a Soldered Joint

The contact area between the conductors is increased slightly by the solder metal. However, you must not rely on the solder to provide a path for electrons. The current must flow through the contact areas, directly **from copper to copper.** Therefore, a joint must be well made **before** it is soldered.

16-2A Heat, Flux and Solder

To solder a joint effectively, you need heat energy, flux, and solder.

The heat energy provided by the soldering iron must raise the temperature of the entire joint **above** the melting point of the solder alloy as quickly and efficiently as possible.

When the temperature of copper is raised, black copper oxide is formed. The molten solder cannot dissolve this oxide and is rejected by it. An oxidized joint cannot be soldered.

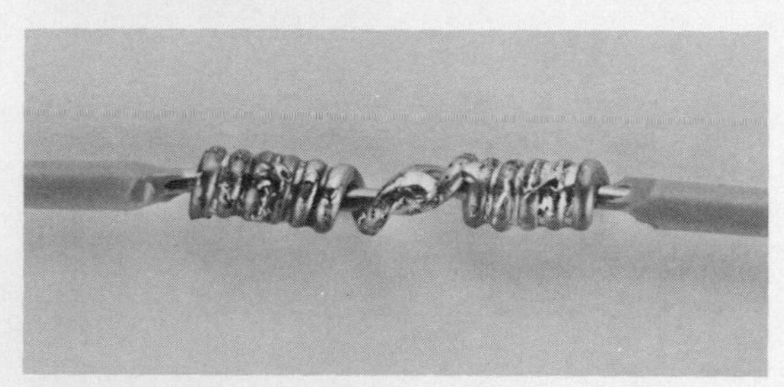

Figure 16-14 Joint Oxidized During Soldering

Flux is a chemical which excludes the **oxygen** during the soldering process and cleanses the copper surfaces of microscopic particles; but it cannot remove dirt. For this reason, a joint must be thoroughly clean before soldering is attempted.

Although various fluxes are available for soldering, only **rosin flux** should be used by the beginner. Rosin is a hard and brittle resin of amber colour which melts at 248°F (120°C). It is available in three forms: as solid blocks that have to be melted with the soldering iron, as soldering paste, and as a continuous-core inside solder alloy wire.

When rosin-core solder wire is used in soldering, the heat melts the rosin **before** it melts the solder. The rosin spreads through the entire joint, followed quickly by the molten solder.

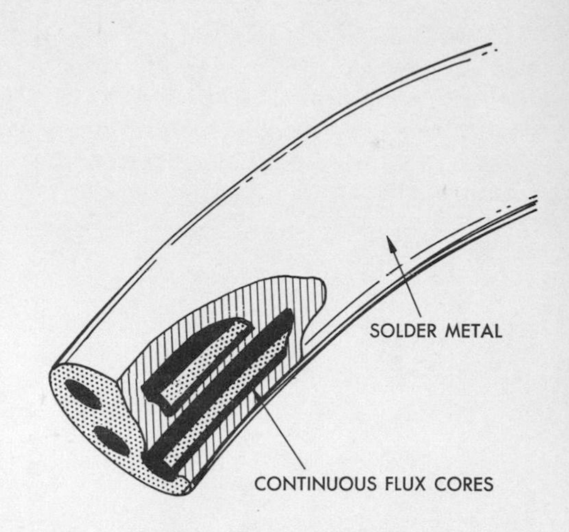

Figure 16-15 Continuous Flux Cores in Solder Wire

The **solder alloys** commonly used in the electrical shop are composed of tin and lead. The tin/lead ratio usually is stamped on the solder bar or printed on the spool containing the solder wire.

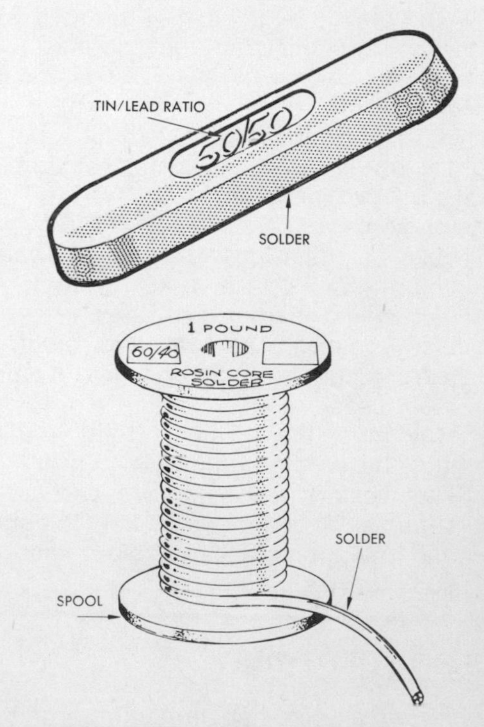

Figure 16-16 Various Forms of Solder Alloy

Solder alloys have tin/lead ratios ranging from 10/90 (softest) to 90/10 (hardest). The 60/40 alloy has unique properties which make it most useful for electrical and electronic work. It has a very low melting point (370°F or 188°C), great mechanical strength, and freezes more quickly than other tin/lead combinations.

For soldering joints, 60/40 wire solder with a continuous core of rosin flux is highly recommended.

16-2B **Preparations for Soldering**

Although flame-heated soldering coppers must still be used on construction sites, the electrical soldering iron has become the common soldering tool in the electrical shop.

The main part of the soldering iron is its copper heating tip. The flat surfaces of the tip are used for **transferring heat** to the joint to be soldered. These surfaces must be in excellent condition, free from pitting and oxidation.

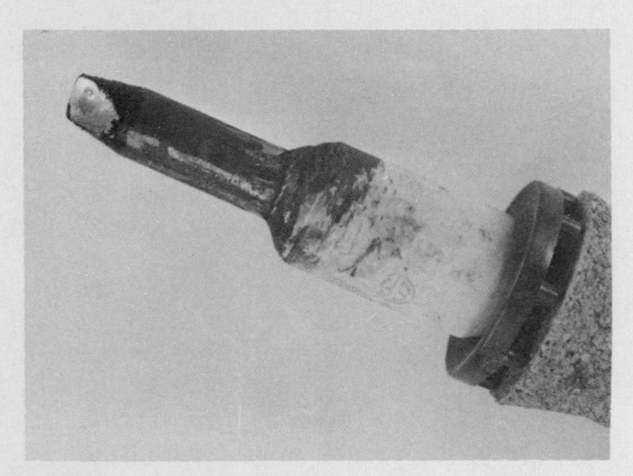

Figure 16-17 Heating Tip of Soldering Iron

Many new soldering irons have plated heating tips to prevent pitting and erosion. Such tips must never be filed. Keep them clean by wiping them carefully with an old rag when the tip is at soldering temperature.

New tips, or new surfaces on old heating tips, must be tinned before they can be used. If a plated tip is not tinned when heated for the **first time,** it will quickly become oxidized and useless for soldering.

To tin the tip, wait until it has reached soldering temperature. **At this instant,** cover the entire tip with a thin coat of flux and solder, and wipe off the excess with a rag. The iron is now ready for use.

16-2C **Soldering a Joint**

Solder as many joints as necessary to perfect your soldering technique. Show each completed joint to your teacher before you solder the next one.

Materials:

One clean, well-made joint for each soldering job

a soldering iron with stand

a length of wire solder, 60/40, with rosin-flux core

a piece of fibreboard to protect your desk

a pair of heavy pliers, or a weight

Procedure:

Place the joint on the fibreboard and bend the conductors upward as shown in Figure 16-18. Using the pliers as a weight, keep the joint about 4 inches (10 cm.) above the board to free your hands for the soldering operation.

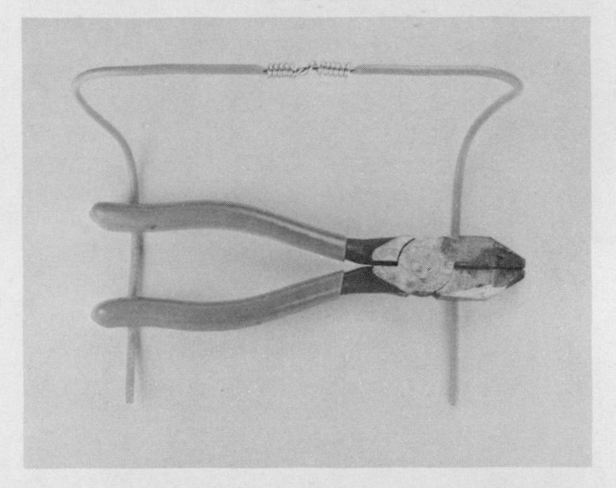

Figure 16-18 Preparing the Joint for Soldering

Put the soldering iron on its stand, plug its linecord into an outlet, and wait until its tip has reached operating temperature.

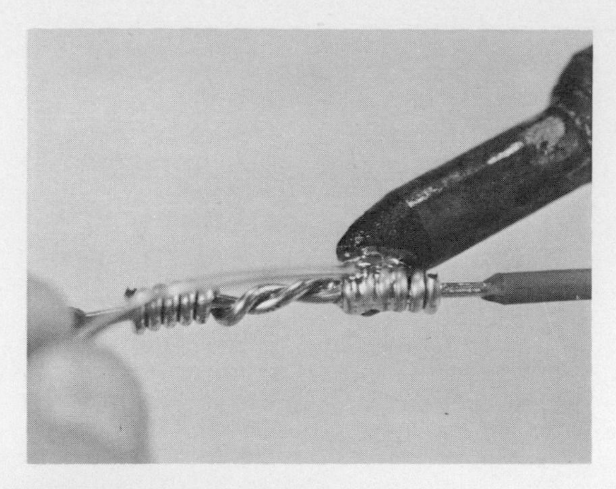

Figure 16-19 Heat Transfer Through a Solder Bridge

Now, touch the underside of the joint with the flat surface of the heating tip. At the same instant, stab the solder wire into the crevice between the tip and joint to form a **solder bridge.** The purpose of this bridge is to conduct heat and flux into the joint.

Now, while holding the heating tip with its solder bridge to the underside of the joint, feed in more flux and solder **from the top.** Gravity and adhesion will cause the flux and solder to reach even the innermost parts of the joint. As the solder spreads, move the wire solder slowly along the topside of the joint, following behind with the heating tip on the underside. Do not remove the tip and solder until all copper surfaces are covered with a thin, even coat of solder.

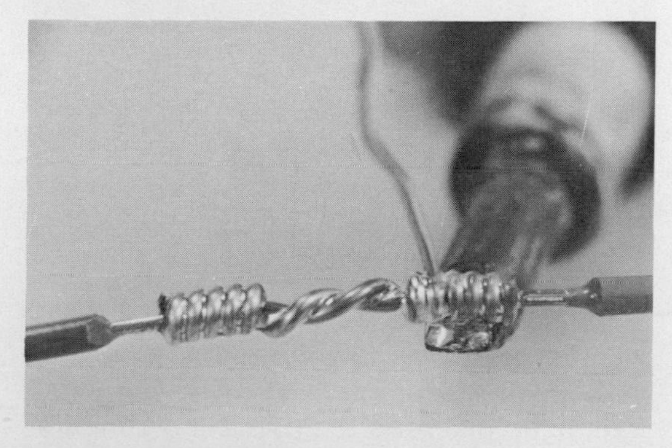

Figure 16-20 Sweating in the Solder

Some excess solder usually appears along the underside of the joint. To remove this, run the heated tip along the underside of the finished joint. The solder will cling to the tip and can be wiped off with a rag.

Do not move the joint or blow at it while the solder is freezing. A joint which has been damaged in this way has a rough, dull appearance; it must be resoldered.

An expertly soldered joint should be covered entirely with a thin coat of glossy solder. The individual turns of the copper wire should still be recognizable, but no bare copper should be left inside the joint.

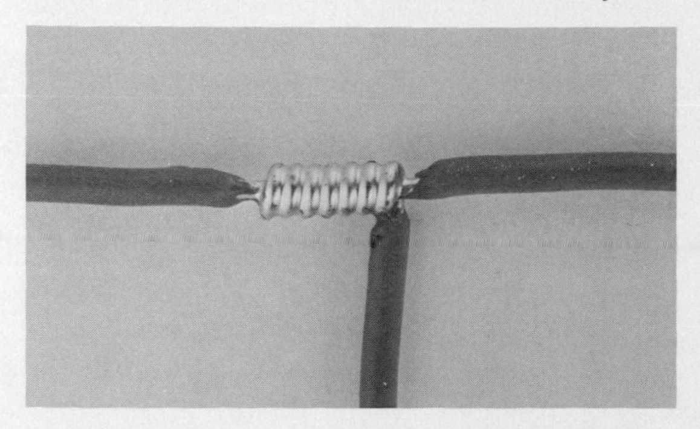

Figure 16-21 Close-Up of Well-Soldered Tap Joint

Assignments:

1. Describe the physical appearance of a well-soldered joint.
2. Explain the purpose of the solder bridge.
3. Why must the flux spread quickly to all areas before the joint is hot enough to melt the solder?
4. What are the reasons for heating the joint from below while feeding in the solder at the top?
5. Why is the cooling period the most critical time in soldering a joint?

16-3 INSULATING CONDUCTOR JOINTS

The Electrical Safety Code rules that a hand-made and soldered joint or splice "shall be covered with insulation approved for the purpose." In modern practice, and for circuits working with voltages below 250V, this usually means an overlapping layer of vinyl insulating tape (#33).

You should insulate several soldered joints to learn proper insulating technique. Show each completed joint to your teacher before you do the next one.

Materials:

one soldered joint for each job
one roll of vinyl insulating tape (#33)
a knife or diagonal cutter

Procedure:

Beginning and ending on the insulation of the conductors, wrap the plastic tape tightly around the entire joint. The turns should overlap by half the width of the tape to produce a double layer of moisture-proof insulation.

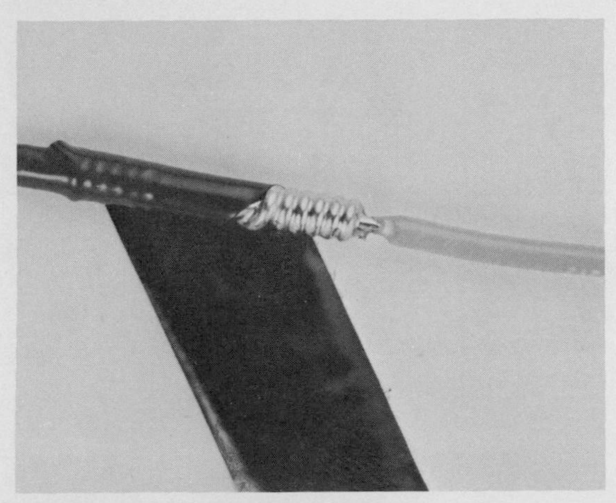

Figure 16-22 Wrapping Vinyl Insulating Tape Around a Western Union Joint

Cut the tape with a knife or diagonal cutters when the joint is insulated. Do not rip it off, since this might alter the shape and thickness of the tape.

Assignments:

1. What is meant by the phrase "insulation approved for the purpose?"
2. Why must you overlap the tape when insulating a joint? (State several reasons.)
3. Vinyl insulating tape cannot be applied at extremely low or extremely high temperatures. What could be the reasons for these limitations?

16-4 USING SOLDERLESS WIRING CONNECTORS

The most common solderless connectors are the set-screw and the twist-on connectors for making rat-tail joints inside fixtures and junction boxes.

16-4A Joining Conductors with a Twist-On Connector

Join several pairs of AWG 14 conductors by using twist-on connectors. Show each completed joint to your teacher before you make the next one.

Materials:

several one-foot (30 cm.) lengths of insulated AWG 14 wire
several twist-on connectors
a knife

Procedure:

Strip about ½ inch (1.2 cm.) of insulation from one end of each conductor. Grasp the two conductors firmly in your left hand, holding the bare ends parallel to each other. Push the connector on the bare ends and twist it onto the conductors in a clockwise direction as far as possible.

The plastic case of the connector should cover the joint so completely that no bare copper remains visible. Tug at the connector to make sure that it is firmly in place.

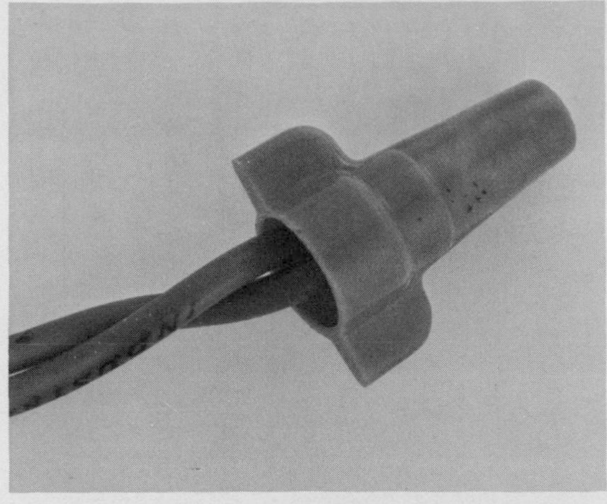

Figure 16-23 Rat-Tail Joint Made with Twist-On Connector

Assignments:

1. Make a neat cross-sectional drawing of a rat-tail joint formed with a twist-on connector.
2. What are the advantages of joining wires with twist-on connectors?
3. Explain why the insulation of the conductors must extend into the plastic cap of the connector.

16-5 MEASURING VOLTAGE, CURRENT AND RESISTANCE

Perhaps the most essential skill in the electrical shop is your ability to use the various meters for measuring voltage, current and resistance. These instruments can be damaged quite easily through a mo-

Figure 16-24 Typical Meter Scale

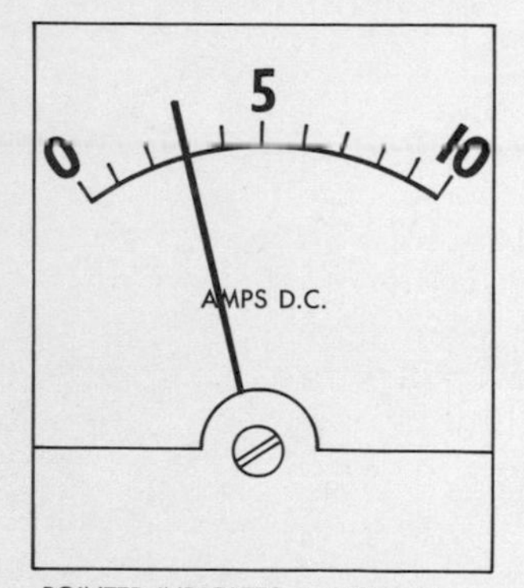

POINTER INDICATES A FLOW OF 3 AMPERES THROUGH THE METER

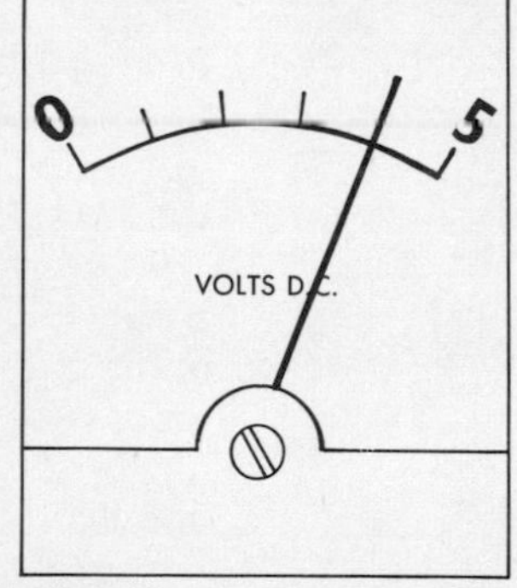

POINTER INDICATES AN E.M.F. OF 4 VOLTS ACROSS THE TERMINALS OF THE METER

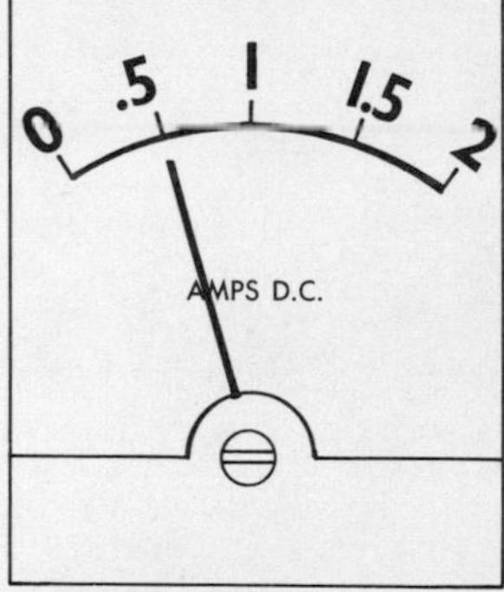

POINTER INDICATES AN ELECTRON CURRENT OF 0.5 AMPERES THROUGH THE METER

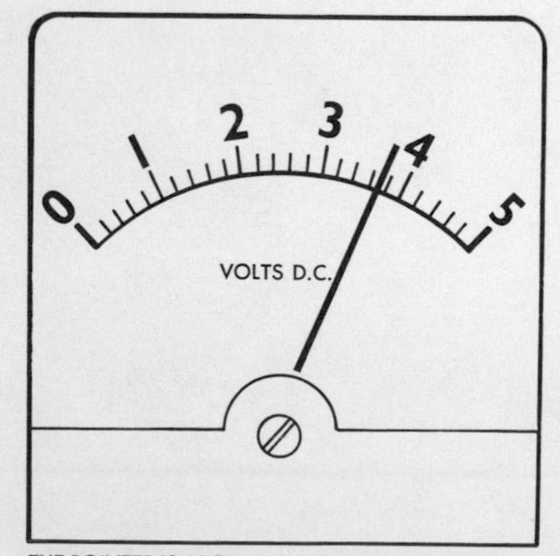

THE POINTER IS ABOUT 7/10 OF THE WAY TOWARD THE 4 VOLT MARK (THE SMALL LINES INDICATE 2/10 OF A VOLT EACH). THE READING IS 3.7 VOLTS

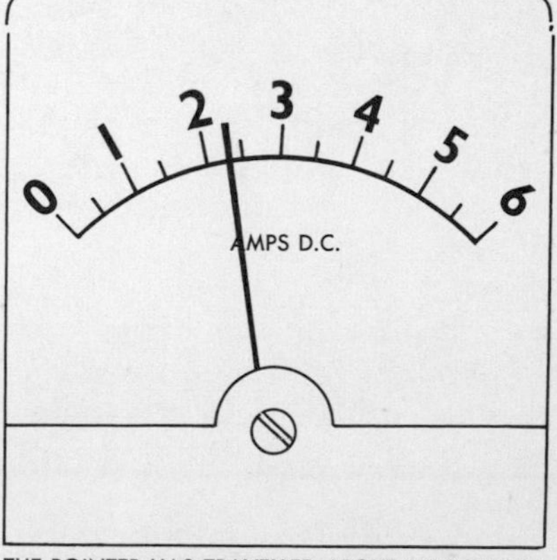

THE POINTER HAS TRAVELLED ABOUT 3/10 OF THE DISTANCE BETWEEN THE 2 AMP. AND 3 AMP. MARK. THE READING IS 2.3 AMPS (APPROX.)

Figure 16-25 Sample Readings on Volt and Ammeter Scales

mentary lapse of caution and they also require considerable skill in the reading of their scales.

The basic techniques for measuring voltage or current or resistance are the same whether you use a single-purpose, single-range meter or a multi-range VOM (Volt-Ohm-Amperemeter). Learning to read the scale of a conventional meter is best accomplished through practice. Study the meter faces shown in Figure 16-25 and compare their pointer positions with the actual measurement value given below each meter.

For technical reasons, many ohmmeters have scales which have their zero points on the right-hand side of the meter face. Their pointers advance from right to left, opposite to the direction of conventional voltmeters and amperemeters.

16-5A Measuring Voltage

Check the scale (and range switch) of the meter to make sure that you have the correct type of instrument for the job. Never attempt to measure a d.c. voltage with an a.c. meter, or vice versa.

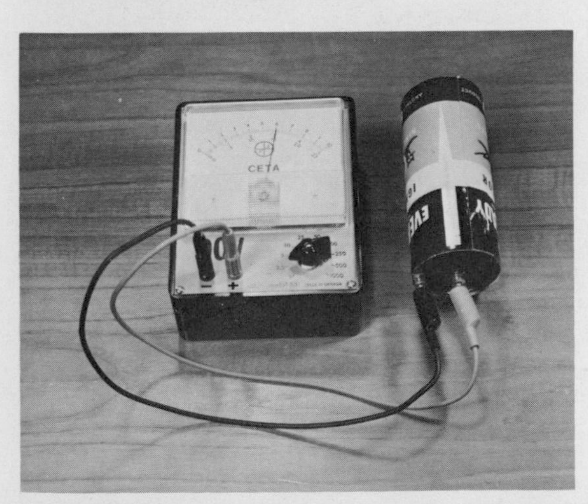

Figure 16-26 Connecting a Voltmeter to a Voltage Source

The two leads of the voltmeter must be connected directly to the terminals of the device whose voltage is to be measured. Connect the negative (black) lead of the meter to the negative terminal of the voltage source, and the positive (red) lead, to the positive terminal.

A voltmeter can measure both applied voltage and voltage drops. Since a voltmeter has an extremely high internal resistance, it draws only a few millionths of an ampere from the circuit to be measured. But in some cases even this small current can "load" or reduce the actual voltage to a lower value.

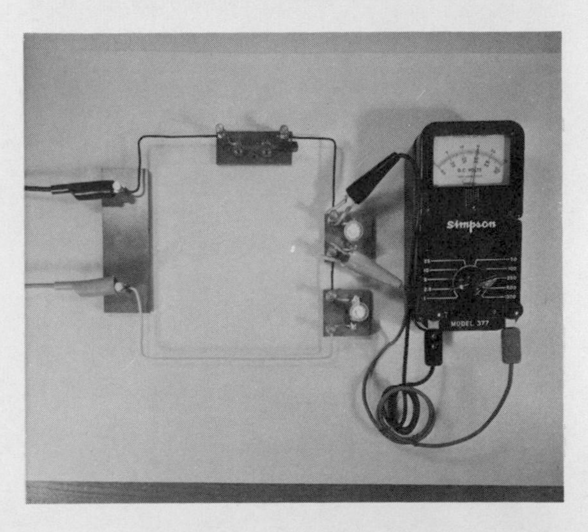

Figure 16-27 Connecting a Voltmeter to Measure a Voltage Drop

Always choose a voltmeter whose full-scale value is higher than the voltage reading you expect from a circuit, or begin with the highest range of a multi-range meter and switch to a lower range until the pointer reaches the middle of the scale.

16-5B Measuring Current Flow

Although an amperemeter looks much like a voltmeter, its internal construction is different. Where a voltmeter draws only a few millionths of an ampere to measure the voltage of a source, an ammeter has the **full circuit current** flowing through its mechanism. As a result, an ammeter can be damaged far more easily than a voltmeter.

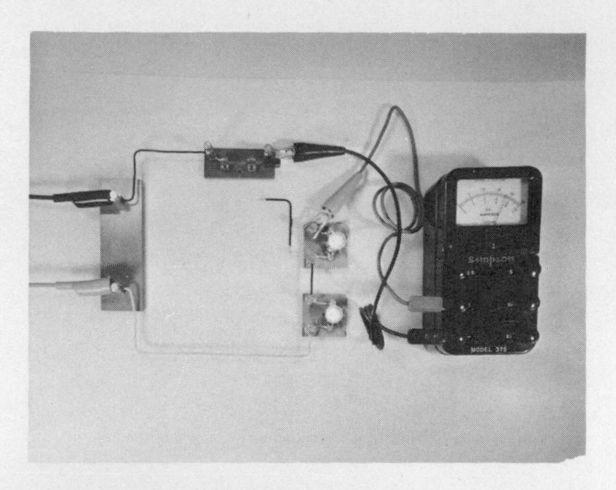

Figure 16-28 Connecting an Ammeter to a Circuit

To measure the current flowing in a circuit, the ammeter must be made a part of the circuit, or connected **in series** with it (Figure 16-28). Before you do this, you must remove the voltage source from the circuit (or turn off the power supply).

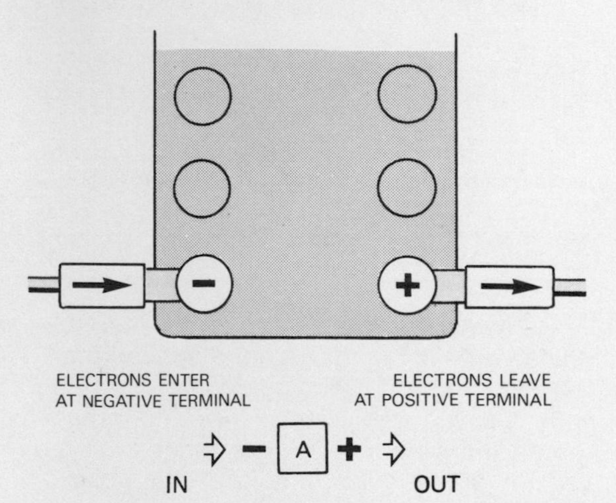

Figure 16-29 Direction of Electron Flow Through a D.C. Ammeter

In a d.c. ammeter, the electrons must enter the meter at the negative terminal and leave it at the positive terminal. If this direction is reversed, the pointer will deflect backwards and the meter may be damaged.

Never connect an ammeter directly to the terminals of a voltage source. The meter's internal resistance is so low that such a connection would be a **short circuit** that could ruin both the voltage source and the ammeter.

Always select an ammeter whose measuring range exceeds the current levels you expect to measure, or begin with the highest range of a multi-range meter and switch to a lower range until the pointer reaches the middle of the scale.

16-5C Measuring Resistance

Voltmeters and ammeters receive the electric energy for moving their pointers from the circuits to which they are connected. This is not so in the case of resistance-measuring devices, or ohmmeters. All ohmmeters have **internal batteries** which supply the measuring current (see Fig. 16-30).

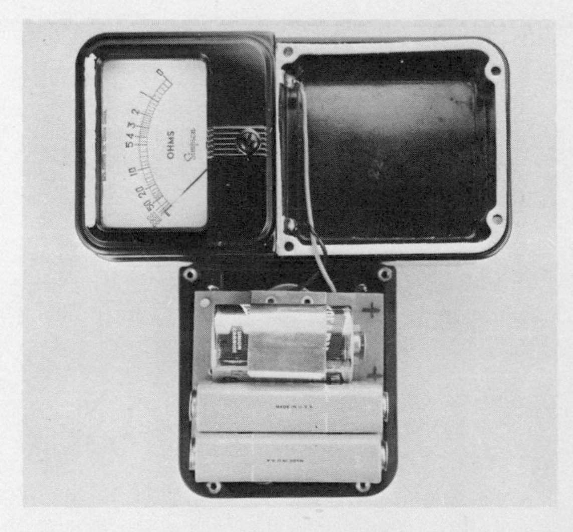

Figure 16-30 Inside View of an Ohmmeter

To measure the resistance of an electrical device, you must **remove it from its circuit.**

Turn the ohmmeter on and short-circuit its leads. (This means zero resistance between their terminals.) Rotate the meter's zero adjust control until the pointer indicates zero ohms. This "calibrates" the meter, making it ready to take measurements.

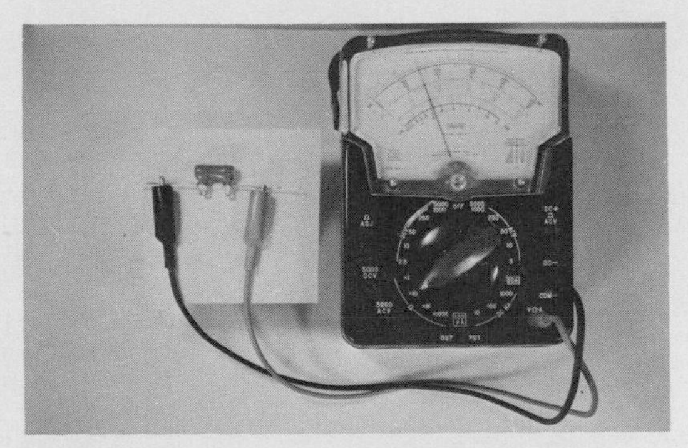

Figure 16-31 Measuring the Resistance of a Wire-Wound Resistor

Now separate the leads again and connect them to the terminals of the device to be measured (Figure 16-31). The pointer will move from right to left and indicate the resistance of the device on the meter's scale.

16-5D **Using Multi-Range Meters**

Figure 16-32 Multi-Range Voltmeter

Multi-range voltmeters and ammeters are far more useful than single-range meters, because they allow a more accurate reading of voltage and current values. The same rules apply as for the use of single-range meters of the same type.

When using a multi-range meter, always set the meter's range selector switch to the **highest range,** then switch to successively lower ranges until the pointer moves to approximately mid-scale. At that point, take the reading.

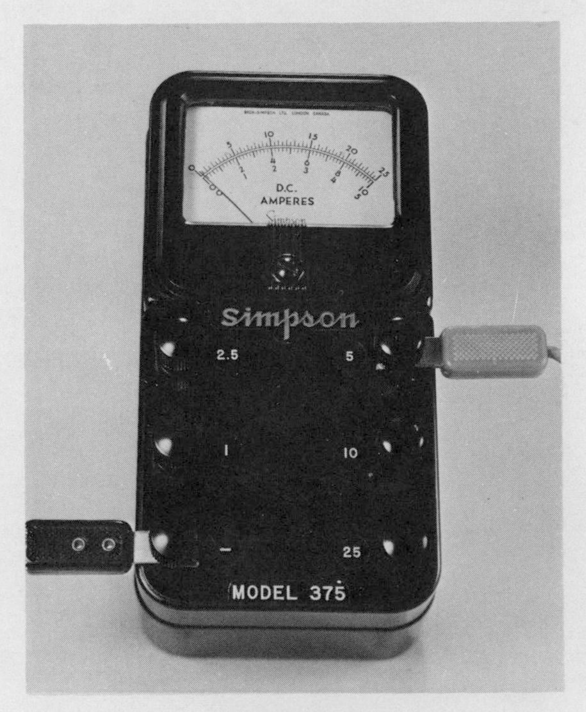

Figure 16-33 Multi-Range Ammeter

17

EXPERIMENTS WITH MAGNETISM

Before you begin with the work of this chapter, here is a brief outline on how to conduct successful experiments:

> Make certain that you have a clear idea of what it is you want to investigate.
> Obtain all necessary materials before you begin your experiment and check them over carefully.
> Follow the instructions as closely as possible and conduct each experiment one step at a time.
> Get involved with what you are doing by concentrating on your work.
> Write a brief but accurate note on the experiment.
> Do the assignments accurately and faithfully. They will help you to remember what you have learned.

17-1 LOCATING THE POLES OF A BAR MAGNET

Materials:
one bar magnet
some coarse iron filings
a sheet of foolscap

Procedure:
Put the sheet of foolscap on your desk and pour the iron filings on it, forming a small heap. Dip first one end of the magnet and then the other into the filings. Observe carefully how the filings are aligned by the magnetic force. Make a rough sketch showing the results of the experiment.

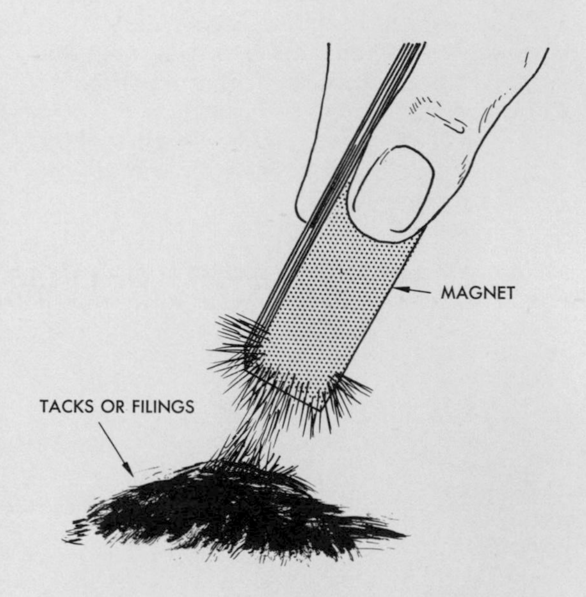

Figure 17-1 Locating the Poles of a Bar Magnet

Assignments:
1. Make a neat drawing showing the alignment of the iron filings at the ends of the magnet.
2. Draw thin lines in the direction indicated by the filings and mark the point where those lines flow together on the magnet. (These are the poles.)
3. From the results of your experiment, explain the meaning of the following statement: the force of a magnet is strongest at its poles.

17-2 IDENTIFYING THE POLES OF A MAGNET

Materials:
one bar magnet
a piece of string, 2 feet (60 cm.) long
a piece of chalk

Procedure:
Tie the magnet to the string and let it hang free above your desk, well away from any steel objects. Wait until the magnet has stopped moving. Take the chalk and mark a capital N on that end of the magnet which points toward the north pole of the earth. Mark the opposite end with a capital S. (If the north direction in your classroom is unknown to you, find it with a magnetic compass before you start the experiment.)

Assignments:
1. Explain why the term "northseeking pole" has been chosen for the N-pole of a magnet.
2. What do you think the first use was of the northseeking property of magnets?

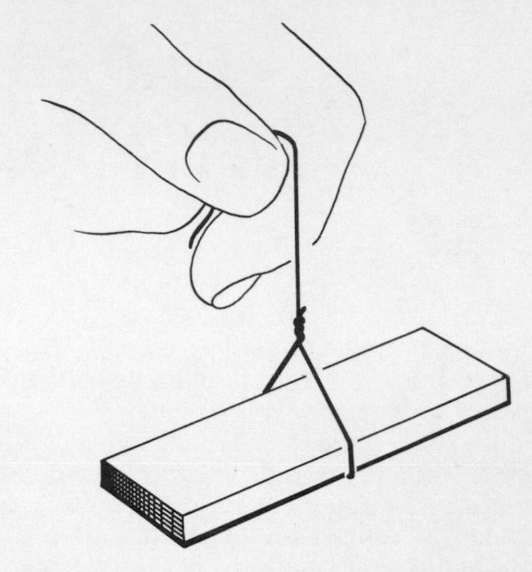

Figure 17-2 Identifying the Poles of a Magnet

17-3 SHOWING THE PATH OF LINES OF MAGNETIC FORCE

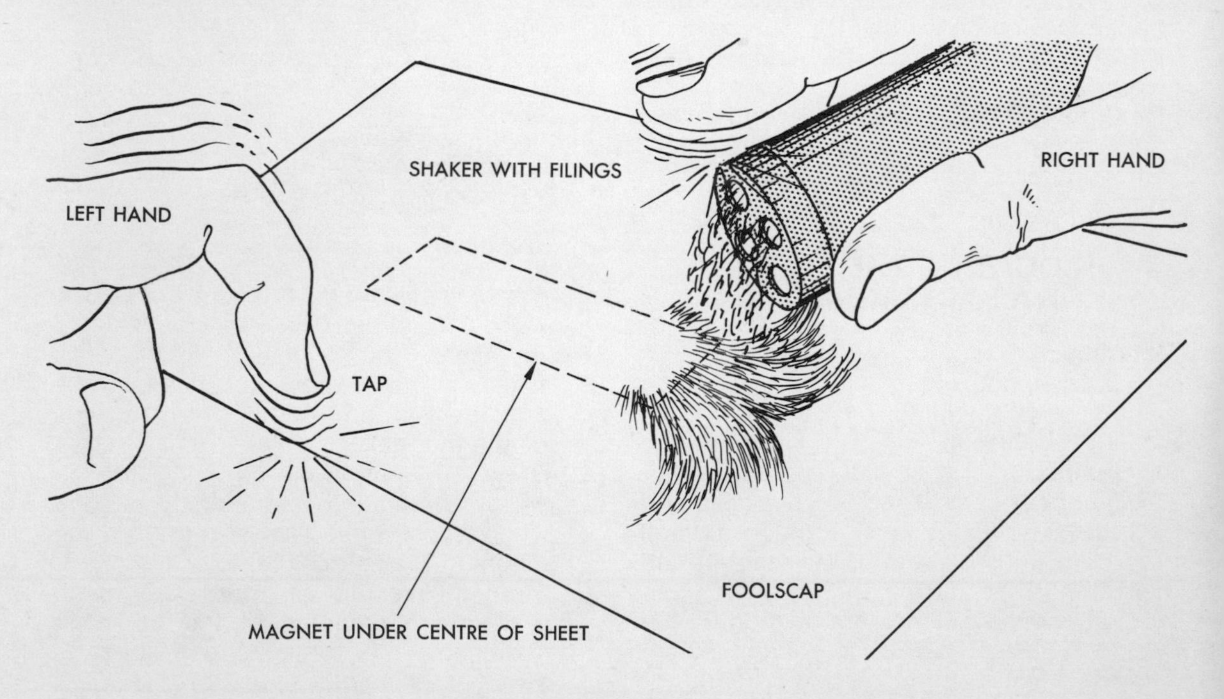

Figure 17-3 Showing the Path of Lines of Magnetic Force

Materials:
one bar magnet
a shaker with fine iron filings
a sheet of foolscap

Procedure:
Place the bar magnet on your desk and put the sheet of foolscap on it. Make certain that the magnet lies under the centre of the sheet. Sprinkle the iron filings evenly over the entire surface of the foolscap by tapping the shaker lightly with your finger. A thin layer of filings works best. Tap the foolscap with your finger to help the filings move against the friction of the surface. Make a drawing of the pattern formed by the iron filings. Then lift the paper carefully off the magnet and pour the filings back into the shaker.

Assignments:
1. Make a neat drawing of the iron-filings pattern obtained in the experiment. Show the outline of the magnet with broken lines.
2. Draw a lines-of-magnetic-force map based on the iron-filings pattern.
3. Describe the appearance of the lines-of-force pattern and state the characteristics of lines of magnetic force revealed by the experiment.

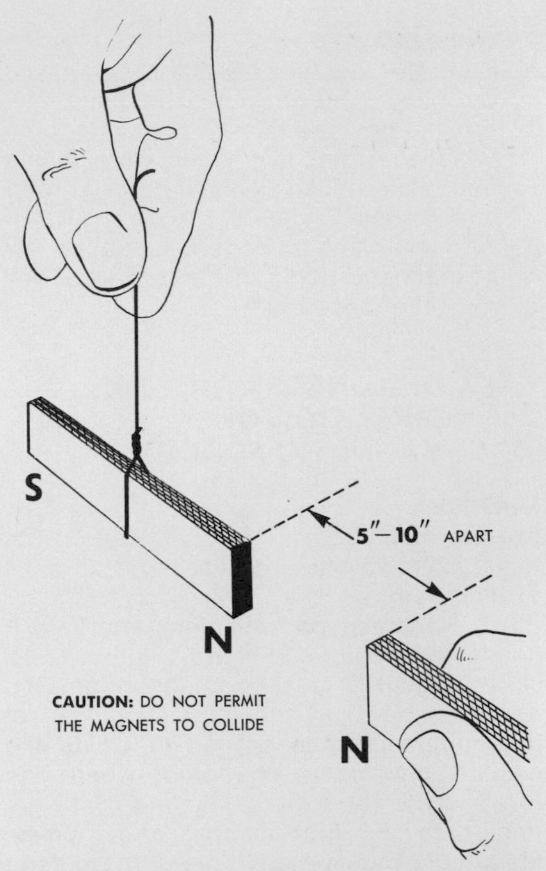

Figure 17-4 Proving the Law of Magnetic Poles

17-4 PROVING THE LAW OF MAGNETIC POLES

Materials:
two identical bar magnets
one 3-foot (90-cm.) piece of string
iron filings in a shaker
a sheet of foolscap
a flat eraser

Procedure:
Tie the string around the centre of the magnet, making sure that the magnet is well balanced and cannot slip out. Let the magnet hang freely above your desk. As soon as it has come to rest, pick up the second magnet and bring its N-pole close to the N-pole of the suspended magnet. Observe what happens and record the results.

Repeat the experiment, but this time bring the S-pole of the second magnet close to the N-pole of the suspended magnet. Be careful. If the magnets are powerful, do not bring them so close that they clash together. (This would reduce their strength considerably.) Again record your observations.

Next, place the 2 magnets on your desk, with their N-poles facing, but separated by the flat eraser. Put the foolscap on top of the magnets and sprinkle a thin, even layer of iron filings on it. Lightly tap one corner of the paper to assist the magnetic force in aligning the filings. Draw a neat map of the iron-filings pattern. Then return the filings into the shaker.

Repeat the last part of the experiment, but with **unlike** poles facing. Again draw a map of the resulting iron-filings pattern.

Assignments:

1. State the law of magnetic poles, based on the results of your experiment.
2. Explain how the lines of magnetic force behaved when like poles were facing and when unlike poles were in front of each other.
3. Can you name an application of the law of magnetic poles in the design of any household objects?

17-5 DEMONSTRATING THE DIRECTION OF MAGNETIC FORCE

Materials:
two bar magnets
ten small magnetic compasses

Procedures:
Place 1 magnet on your desk and hold it down **firmly** with 2 fingers (Figure 17-5). Now, slide the second magnet into the position shown in the illustration and let the magnetic force act on it. Guide the second magnet to the point where the force wants to move it, keeping it at right angles to the first magnet at all times. Make sure to use the N-pole of the second magnet. Repeat the experiment several times and record your observations.

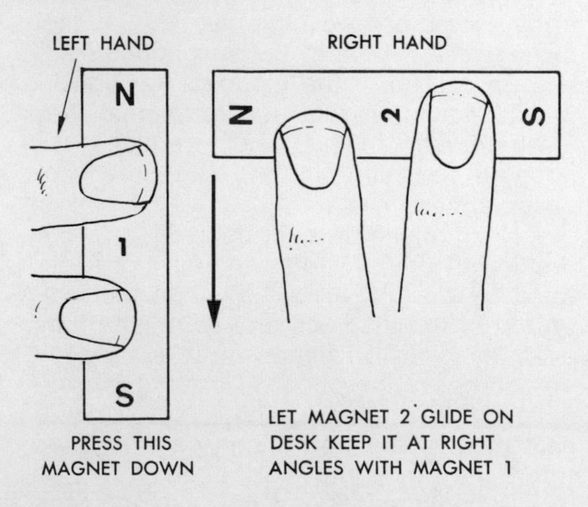

Figure 17-5 Demonstrating the Direction of Magnetic Force with Bar Magnets

Next, put 1 bar magnet in the centre of your desk (well away from the other magnet) and place the 10 magnetic compasses around it in the manner shown in Figure 17-6. Make a neat drawing of the arrangement, showing clearly the alignment of the compass needles. Colour their N-poles to emphasize the direction of the alignment.

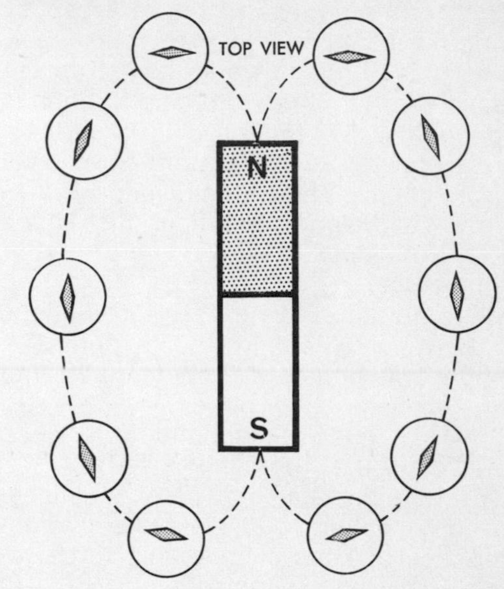

Figure 17-6 Demonstrating the Direction of Magnetic Force with Small Compasses

Assignments:

1. State the general rule about direction of magnetic force.
2. Why is it proper to say that magnetic force "acts" in a certain direction, rather than "flows" or "moves?"
3. Why is the international agreement regarding the direction of magnetic force an arbitrary choice?

17-6 IDENTIFYING MAGNETIC AND NON-MAGNETIC MATERIALS

Materials:
one bar magnet
small sheets of iron, nickel, aluminum, glass, plastic and other materials

Procedure:
Place the material samples on your desk and try to pick them up with the magnet, one by one. Separate the samples into magnetic and non-magnetic materials and record the results.

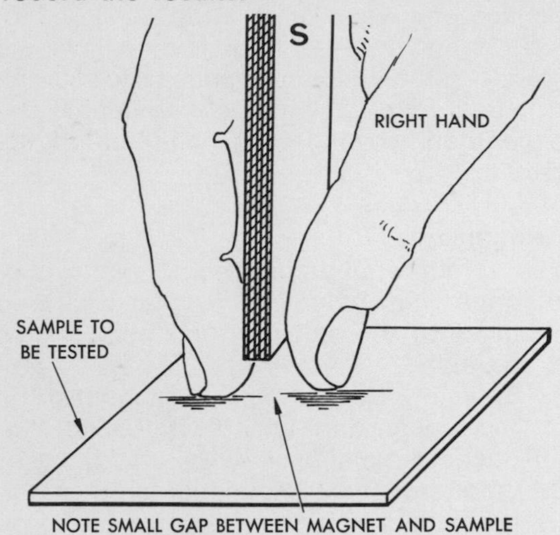

Figure 17-7 Identifying Magnetic and Non-Magnetic Materials

Assignments:
1. Write a neat list of each group of materials, magnetic and non-magnetic.
2. What do all magnetic materials have in common?

17-7 INVESTIGATING FLUX LINES IN VARIOUS MATERIALS

Materials:
one horseshoe magnet with keeper
a piece of copper or aluminum, same size as keeper
a flat eraser
a sheet of foolscap
a shaker with iron filings

Procedure:
Put the horseshoe magnet on your desk and place the piece of copper or aluminum in front of its poles, separated by the eraser (Figure 17-8). Cover the whole set-up with the sheet of foolscap and sprinkle a thin layer of iron filings on it. Tap one corner of the paper to assist the magnetic force in aligning the filings. Draw an accurate map of the filings pattern. Then return the filings to the shaker.

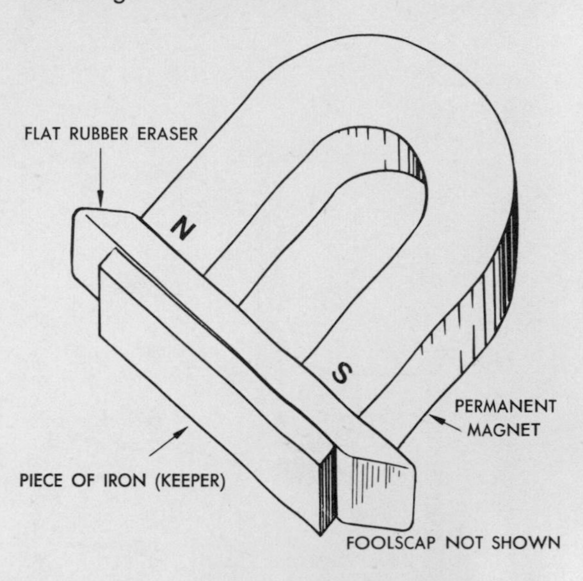

Figure 17-8 Set-up for Investigating Flux Lines

Next, arrange the horseshoe magnet and eraser as before, but with the iron keeper in place of the copper or aluminum piece.

Make an iron-filings pattern and draw a map showing the paths of the flux lines.

Assignments:
1. Draw a neat lines-of-force map showing the effect of a non-magnetic material on flux lines.
2. Draw a second lines-of-force map showing the effect of a magnetic material on flux lines.
3. Explain how flux lines behave in a) magnetic materials b) non-magnetic materials.

17-8 INDUCING TEMPORARY MAGNETISM IN SOFT IRON

Materials:
one bar magnet
a piece of soft iron, approximately ¼ x ¼ x 1 inch (.6 x .6 x 2.5 cm.)
a tack, nail, or paper clip

Procedure:

Put the nail or paper clip on your desk and try to pick it up with the piece of soft iron. (Pure soft iron should not attract the clip at all.) Now, while holding the soft-iron piece against the clip or nail with one hand, bring the magnet to within 1/4 inch (.6 cm.) of the soft iron, as shown in Figure 17-9. (The magnet must not touch the iron.)

Begin to lift both the iron and the magnet slowly upwards, without changing the distance between them. Stop about 1 foot (30 cm.) above the desk and remove the magnet. Observe what happens to the nail or paper clip at that instant. Repeat the experiment several times and record the results.

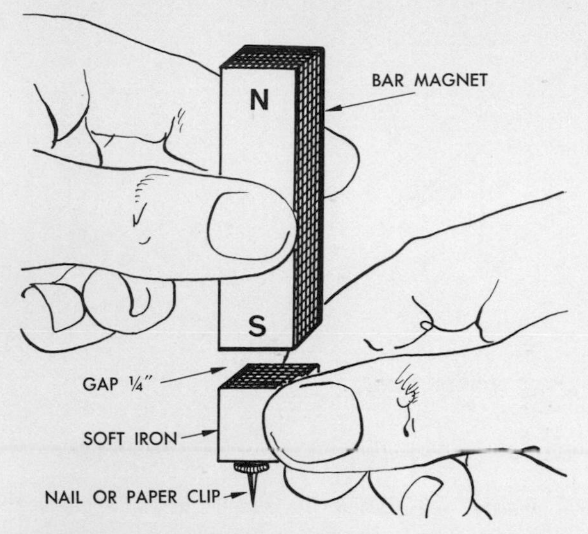

Figure 17-9 Inducing Temporary Magnetism in Soft Iron

Assignments:

1. Explain what happened to the nail or paper clip when the magnet was held close to the soft iron and when it was removed.
2. Explain why this type of magnetism is temporary and why it is called "induced magnetism."
3. What does the magnetic force of the permanent magnet do to the iron atoms?
4. Why is soft iron always attracted to a magnet but never repelled?

17-9 INDUCING PERMANENT MAGNETISM IN STEEL

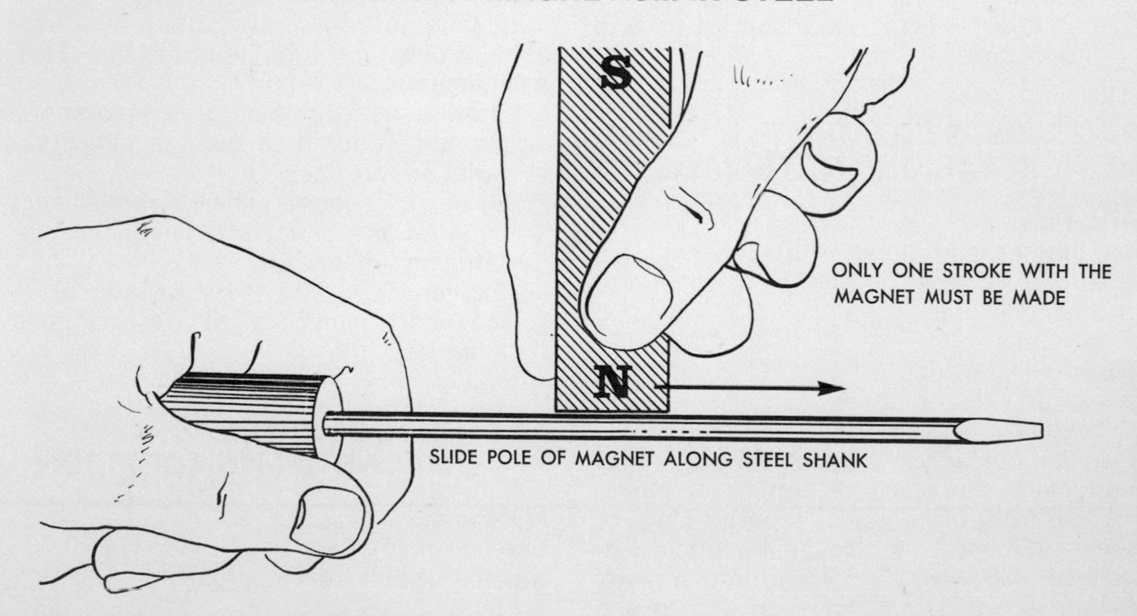

Figure 17-10 Inducing Permanent Magnetism in Steel

Materials:
one bar magnet
one steel-shank screwdriver
a small nail or paper clip
Procedure:
Put the nail on your desk and try to pick it up with the blade of the screwdriver. (If the screwdriver picks up the nail, it must first be de-magnetized by hitting its shank repeatedly with a pair of pliers or with another screwdriver.)

Now, carefully attach the N-pole of the permanent magnet to the screwdriver shank (close to the handle) and slide it slowly along the shank (Figure 17-10). Use only a **single stroke** to magnetize the screwdriver. Once again, try to lift up the nail with the screwdriver's blade. Record your observations.
Assignments:
1. Describe briefly the single-stroke method of magnetizing a steel object.
2. What would happen inside the steel shank if the magnet was stroked along the shank a second time?
3. State the basic difference between soft iron and hard steel with reference to their magnetic properties.
4. Why does steel not lose its magnetism when the magnetizing force is removed?
5. Explain what happens inside the shank of a magnetized screwdriver when you de-magnetize it by hitting it with a steel object.

17-10 TESTING VARIOUS MATERIALS AS MAGNETIC SHIELDS

Materials:
one bar magnet
several thumb tacks
5 x 5 inch (12.5 x 12.5 cm.) sheets of steel, brass, glass, copper, tin, stainless steel, aluminum, and others
Procedure:
Scatter the thumb tacks over a small area of your desk. Then, one at a time, hold the various materials about ¼ inch (.6 cm.) above the clips and place the magnet directly above each sheet. Observe the effect of the magnetic force on the thumb tacks for each material and record your findings.

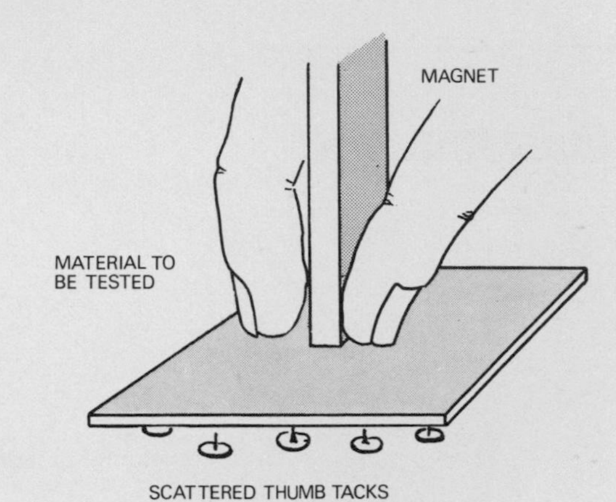

Figure 17-11 Testing Various Metals as Magnetic Shields

Assignments:
1. Divide the materials which you have tested into two groups: "magnetic shields" and "no shielding effect."
2. What happens to lines of magnetic force inside a magnetic shield?
3. Which material would make the best shield? Why?

17-11 SHIELDING A MAGNETIC COMPASS FROM MAGNETIC FORCE

Materials:
one bar magnet
a soft-iron ring, ¼ inch (.6 cm.) thick, 1 inch (2.5 cm.) high, 3 inches (7.5 cm.) in diameter
a magnetic compass
a flat eraser
Procedures:
Put the iron ring on your desk and place the magnetic compass inside of it on top of the flat eraser. Now, bring the N-pole of the bar magnet to within 6 inches (15 cm.)

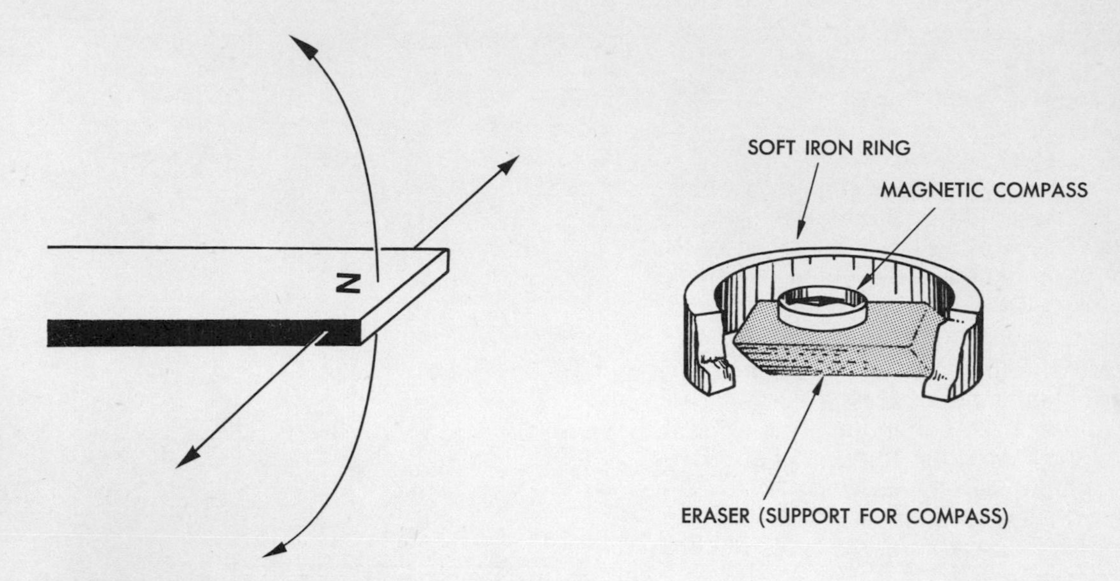

Figure 17-12 Shielding a Magnetic Compass from Magnetic Force

of the iron ring. Move the pole back and forth and watch the reaction of the compass needle. Remove the iron ring and repeat the motions of the magnet, again observing the behaviour of the needle. Write a brief report of the experiment and its results.

Assignments:

1. Draw a lines-of-force map showing how the soft-iron ring protects the compass needle from the force of the magnet.
2. Why is the shielding effect of the soft-iron ring not complete?
3. Sensitive instruments are entirely enclosed in soft-iron casings. Explain the reason for this.
4. What would happen if hard steel were used in place of the soft-iron ring?

17-12 PROVING THAT A MAGNET CONSISTS OF MANY SMALL MAGNETS

Materials:

one bar magnet
one old hacksaw blade
iron filings
a sheet of foolscap
two pairs of lineman's pliers
a pair of safety glasses

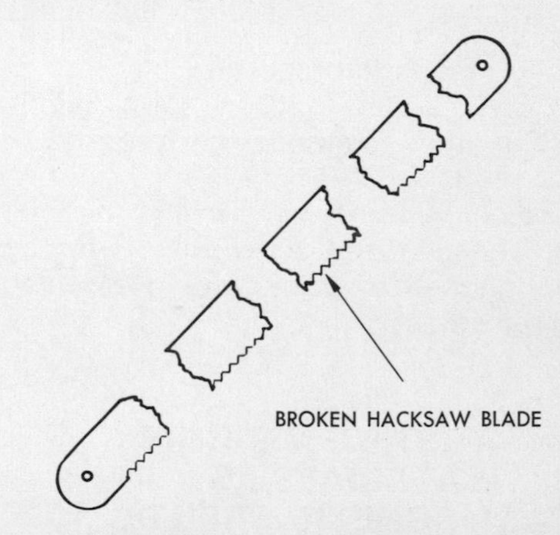

Figure 17-13 Alignment of Broken Hacksaw Blade

Procedure:

Magnetize the hacksaw blade by stroking it with the N-pole of the magnet. A single stroke will do. Next, wearing the safety glasses, take the 2 pairs of pliers and, holding one in each hand, break the hacksaw blade into 6 or 8 pieces. Line up the pieces on your desk exactly in the order in which you broke them off, leaving about ¼ inch (.6 cm.) of space between them. Cover the pieces with the foolscap and make an iron-filings pattern of the lines of force. Draw a lines-of-force map based on the filings pattern.

Assignments:

1. Draw a neat sketch of the iron-filings pattern obtained in the experiment.
2. Since the hacksaw blade was a single magnet before you broke it apart, what conclusions can you draw about the nature of all permanent magnets?
3. Explain what would have happened to the lines of force if you would have broken the hacksaw blade into many more pieces and lined all of them up in the same way as in this experiment.

18

EXPERIMENTS WITH ELECTRICITY

A great deal can be learned from successful experiments. The extent to which your experiments will work and produce the desired results depends on how you approach your work:

> Make certain you have a clear idea of what it is you want to investigate.
> Obtain all necessary materials before you begin your experiment and check them over carefully. Report defective instruments to your teacher.
> Follow the instructions as closely as possible and conduct each experiment one step at a time.
> Get involved with what you are doing by concentrating on your work.
> Read instruments as accurately as possible and record your readings immediately.
> Do the assignments accurately and faithfully; they will help you to remember what you have learned.

18-1 GENERATING ELECTRICITY BY FRICTION

Materials:
one ebonite or plastic rod
a piece of fur
one NE 51 neon lamp with socket

Procedure:
Produce an electric charge on the rod by pulling it quickly through the cat's fur. Ask a fellow student to place a finger on one of the terminals of the lamp socket and touch the other terminal with the charged end of the rod. Observe what happens in the lamp at that instant. Repeat the experiment several times. (It will work better if your classroom is in semi-darkness.)

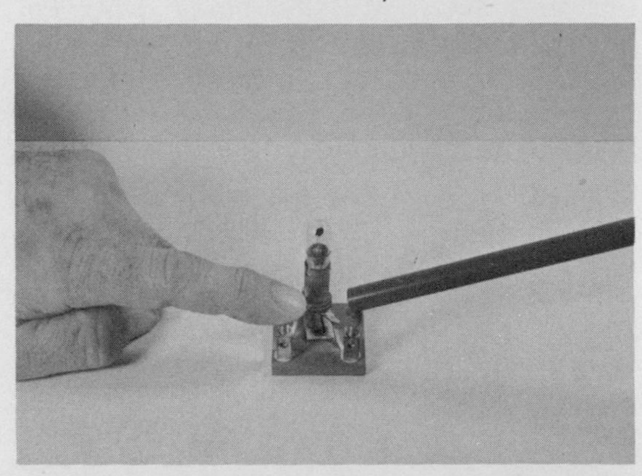

Figure 18-1 Discharging a Plastic Rod Through a Neon Lamp

Assignments:
1. Write a brief account of your observations. Describe the energy conversions that have occurred in the experiment.
2. Compare the amount of electricity generated with the amount of mechanical energy used to produce it. Is this an efficient way of generating electricity?
3. If so little electricity was generated by this method, what happened to the remainder of the mechanical energy?
4. Name and describe briefly an application of static electricity in industry.
5. What is the purpose of having a fellow student touch the lamp's second terminal with his finger?

18-2 GENERATING ELECTRICITY WITH A MOVING MAGNETIC FORCE

Materials:
a bar magnet
25 feet (7.5 metres) of AWG 20 wire, or a coil
a sensitive ammeter (0 to 1 milliampere)
Procedure:
Wind the copper wire into a small coil, 2 to 3 inches (5 to 7.5 cm.) in diameter, and secure it by wrapping short pieces of the wire around the turns. Connect the ammeter to the bared ends of the coil and place the coil flat on your desk, at least 1 foot (30 cm.) away from the meter. Hold the magnet in the centre of the coil and move it slowly up and down, observing the reaction of the meter's pointer at the same time. Write down your observations.

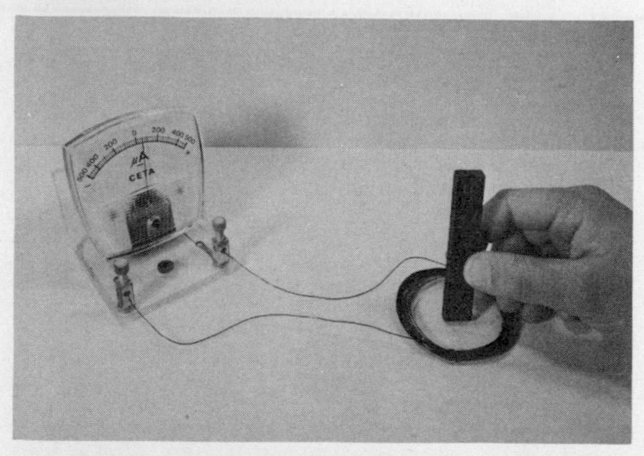

Figure 18-2 Generating Electricity by Electromagnetic Induction

Next, place the magnet on the table, in a standing or vertical position. Pick up the coil and move it up and down, with the magnet standing in its centre. Again observe the reaction of the meter and record your findings.
Assignments:
1. Explain carefully under what conditions electricity was generated and when no energy conversion occurred.
2. Describe the relationship between the speed of motion of the magnet and the reaction of the meter.
3. What happened when you reversed the direction of the magnet's motion?
4. What happened when the magnet was in the centre of the coil, but did not move? (And the coil was also stationary.)
5. Name a device which uses electromagnetic induction and describe briefly how it works.

18-3 GENERATING ELECTRICITY WITH A PIEZOELECTRIC CRYSTAL

Materials:
a piezoelectric crystal in a special holder
a neon lamp NE 51 with socket
connecting wires
a small mallet
Procedure:
Connect the piezoelectric crystal to the terminals of the lampholder. Place the crystal upright on the desk. Darken the room. Hold the crystal at its base and strike a sharp blow on its top with the mallet. Observe the lamp while you do this.

Figure 18-3 Generating Electricity with a Piezoelectric Crystal

Next, using the mallet, put your full weight on the crystal and do not vary the pressure. Again observe the lamp and write down your findings.

Assignments:

1. Explain carefully under what conditions electricity was generated and when no energy conversion occurred.
2. Describe the effect of a steady, unvarying pressure acting on the crystal.
3. Check the meaning of the prefix "piezo" in a dictionary and copy that definition.
4. Name a common use of the piezoelectric effect and describe briefly how the crystal works in the application you have selected.

18-4 GENERATING ELECTRICITY BY CHEMICAL ACTION

Materials:

a small glass beaker
a small quantity of table salt, or some cleansing powder
two strips of copper, roughly 1 x 5 inches (2.5 x 12.5 cm.)
one strip of zinc, same size as copper strips
a voltmeter, 0 to 1 volt d.c., with test leads and alligator clips

Procedure:

Put the table salt (or cleansing powder) into the beaker and half fill the vessel with water. Place the copper strips in this weak acid solution, well apart from each other. Connect the meter leads to the strips, as shown in Figure 18-4. Observe the meter's reaction.

Figure 18-4 Generating Electricity by Chemical Action

Next, remove 1 copper strip, replace it with the zinc and reconnect the meter leads, negative lead to the zinc. Again observe the meter's reaction and record your findings.

Assignments:

1. Explain what happened when a) like metals b) unlike metals were placed in the weak acid solution.
2. Describe the energy conversions which occurred during the experiment.
3. Why must the negative lead of the voltmeter be connected to the zinc strip?
4. In what way does a modern dry cell differ from this reconstruction of the original voltaic cell?
5. When will a primary cell cease to generate electricity?
6. State an alternate name for the weak acid solution in the voltaic cell.

18-5 GENERATING ELECTRICITY BY HEATING A THERMOCOUPLE

Materials:

a thermocouple
an ammeter with leads (0 to 0.1A)
a candle and matches, or a lighter

Procedure:

Connect the ammeter leads to the thermocouple. Hold the couple by one of its leads and place the metal junction (the weld

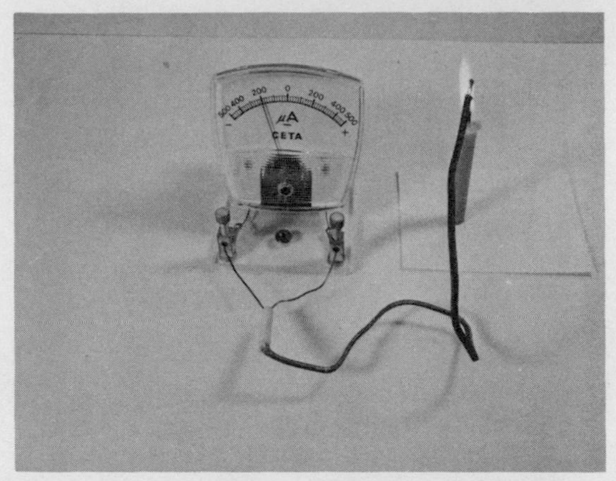

Figure 18-5 Generating Electricity by Heating a Thermocouple

spots) above the flame. Observe the reaction of the meter. (If the needle begins to move in the wrong direction, reverse the meter leads.) Remove the thermocouple from the flame and watch the reaction of the meter as the couple cools down. Write down your observations.

Assignments:

1. Describe the construction of the thermocouple. Make a labelled drawing to illustrate your description.
2. State the results of the experiment. Mention the direction of electron flow in the themocouple-meter circuit.
3. Describe the energy conversion which occurred inside the thermocouple.
4. Explain the relationship between the temperature of the thermocouple and the amount of electricity generated by it.

18-6 GENERATING ELECTRICITY BY SHINING LIGHT ON A SOLAR CELL

Materials:

a mounted solar cell
a voltmeter (0 to 1 volt, with test leads)
a light source

Procedure:

Connect the meter leads to the terminals of the solar cell. Shine light on the cell and observe the reaction of the meter. Interrupt the flow of light energy with your hand, and vary the intensity of the light falling on the cell by moving the light source toward and away from it. Again write down your observations.

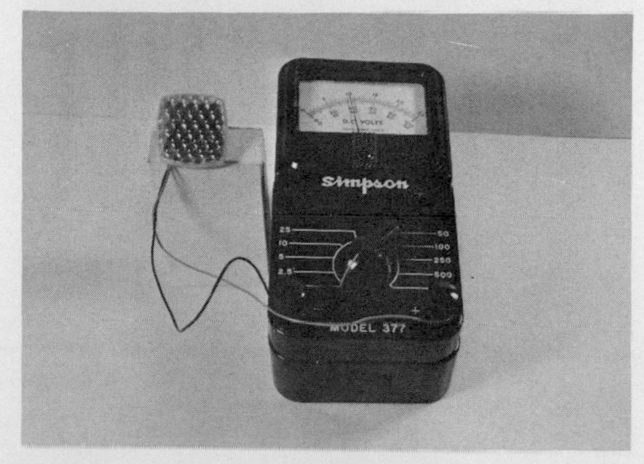

Figure 18-6 Generating Electricity by Shining Light on a Solar Cell

Assignments:

1. Explain the energy conversions that took place in the solar cell.
2. Point out the relationship between the intensity of the light falling on the cell and the voltage generated by the cell.
3. Make a detailed, labelled drawing of a modern photovoltaic cell (solar cell).
4. A modern silicon solar cell converts 14 percent of the light energy falling on it to electricity. What happens to the remaining 86 percent of the light energy?

18-7 CHARGING AND DISCHARGING A SIMPLE SECONDARY CELL

Caution:

The dilute sulphuric acid necessary for this experiment is dangerous. If you should get some of it on your skin, clothes or equipment, rinse it away immediately. Also, handle the lead strips with caution and wash your hands when you have completed the experiment. The room must be well ventilated, because flammable hydrogen gas will be produced while the cell is being charged.

Materials:

a small glass jar
two lead strips, approximately 1 x 4 inches (2.5 x 10 cm.)
a voltmeter, 0 to 5 volts, with leads and alligator clips
a low-voltage d.c. power supply, with leads and alligator clips
a 2.2-volt flashlight bulb with socket and leads
dilute sulphuric acid

Procedure:

Fill the jar half full with dilute sulphuric acid and carefully place the lead strips in the acid, well apart from each other. Connect the voltmeter to the strips and take a reading. Record your findings.

To charge the secondary cell, disconnect the voltmeter from the strips and connect the power supply leads to them. The power supply must be turned **off** and its voltage control at **zero.** Again make sure that the lead strips do not touch each other in the acid solution.

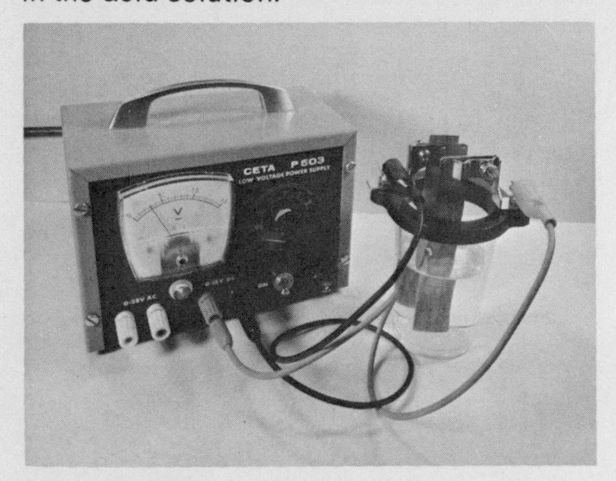

Figure 18-7 Charging a Simple Secondary Cell

Now turn the power supply on and raise its output voltage slowly to 3 volts. Charge the cell for about 3 to 5 minutes. Observe the actions taking place at the electrodes and in the acid solution (colour changes, bubbling, etc.) and write down your observations.

Turn the power supply off and disconnect its leads from the lead strips as carefully as possible.

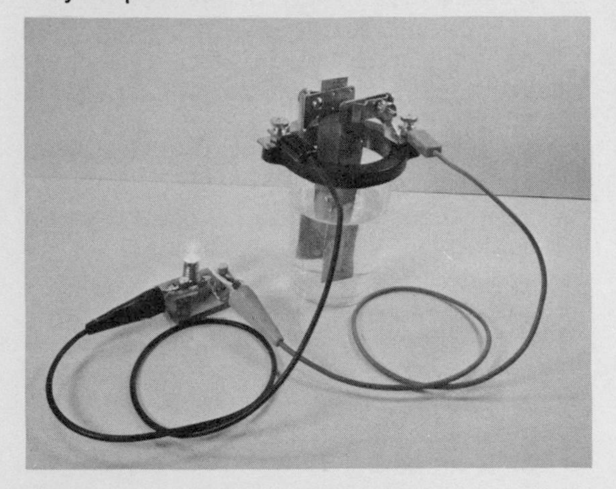

Figure 18-8 Discharging a Simple Secondary Cell

Connect the test leads of the voltmeter to the charged cell and measure its output voltage. Record that value. Remove the voltmeter leads and connect the flashlight bulb to the strips as carefully as possible. Leave the lamp connected until its brilliance decreases and fades out.

Disconnect the lamp from the cell. Remove the lead strips from the beaker in a sink and pour the acid solution carefully down the drain. Wash and dry all parts and return them to storage.

Assignments:

1. Describe how a simple secondary cell is charged and discharged. Make simple drawings to illustrate your answers.
2. State the actions (and changes) which occur at each lead strip when the cell is charged and discharged.
3. Explain why the charged storage or secondary cell actually is a voltaic cell.
4. List the basic differences between primary and secondary cells.

18-8 MEASURING THE VOLTAGE OF A SERIES BATTERY

Materials:

three Number 6 dry cells
a voltmeter, 0 to 5 volts d.c., with test leads
three jumper wires

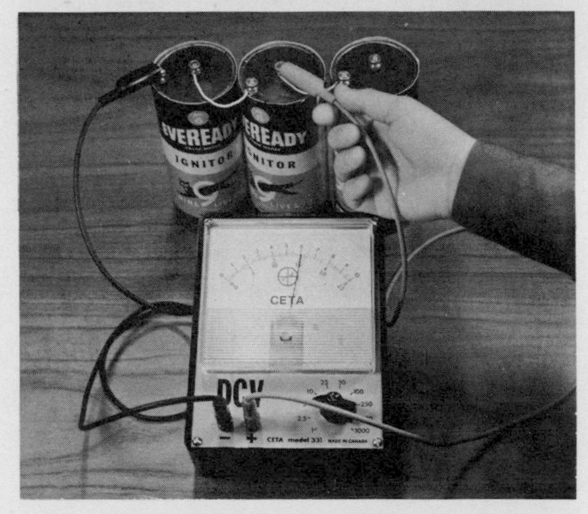

Figure 18-9 Measuring the Voltage of a Series Battery

Procedure:
Place the cells in single file on your desk, all negative terminals to the right. Using the jumper wires, connect the cells in series (— of one cell to + of next cell). Tighten all connections.

Now, moving from cell to cell with the negative meter lead, measure accurately the voltage level at each negative terminal and record these values on a chart.

Assignments:
1. Write a brief account of the experimental set-up, including a schematic diagram of the circuit.
2. Prepare a neat chart showing the voltages measured during the experiment. Use such headings as Cell #1, Cell 1+2, etc.
3. Make a statement comparing the individual cell voltages with the total voltage of the series battery. Express the same statement in algebraic form, using such terms as E_1, E_2, E_T, etc.
4. How many cells would have to be connected in series to produce a) a 450-volt battery? b) a 9-volt transistor battery?

18-9 INVESTIGATING THE RELATIONSHIP BETWEEN VOLTAGE AND CURRENT (OHM'S LAW)

Materials:
a low-voltage power supply, 0 to 10 volts d.c.
a 5-ohm wire-wound resistor
a voltmeter, 0 to 10 volts d.c.
an ammeter, 0 to 12 amperes d.c. (or similar range)
a knife switch (mounted)
a circuit board with terminal posts
insulated copper wire

Procedure:
Construct the experimental circuit, making certain that all connections are properly tightened. Check the completed circuit carefully against the schematic diagram, Figure 18-10. Connect the power supply, but leave its switch in the off position, and its voltage control at zero. Open the knife switch.

Ask your teacher to inspect the circuit. Then, turn on the power supply and close the knife switch. Advance the voltage control **slowly** until the voltmeter **in the circuit** reads exactly 1 volt. Read the ammeter accurately and record the current value. Advance the voltage control to 2, 3, 4 and 5 volts and measure the resulting current each time. Record all readings. Make sure to turn the power off before you dismantle the circuit and return all components to storage.

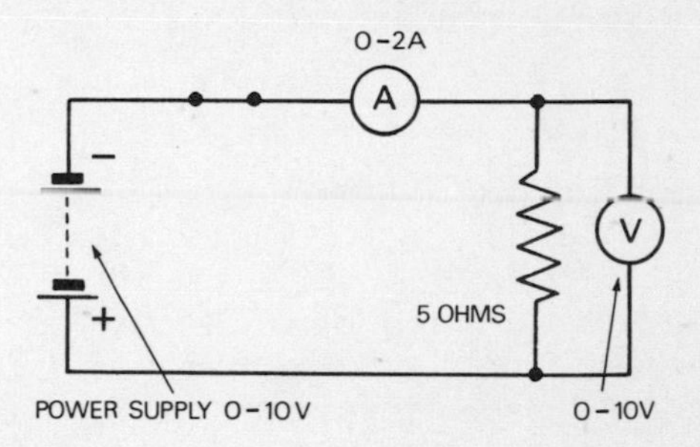

Figure 18-10 Schematic Diagram of Experimental Circuit for Measuring Voltage and Current

Assignments:
1. Make a neat chart of the voltage and current values recorded during the experiment. Use three headlines: Applied Voltage, Resultant Current, and Resistance.
2. Based on the results of the experiment (and ignoring minor inaccuracies in your measurements), explain the following statement: the current flowing in an electric circuit is directly proportional to the applied voltage.
3. Why does an increase in applied voltage result in a greater current flow?
4. A voltage of 10 volts is applied to a given circuit, resulting in a current flow of 5 amperes. How much current will

flow in this circuit if the voltage is a) doubled? b) cut in half? c) reduced to 1 volt? d) increased to 15 volts? e) advanced to 30 volts?

18-10 INVESTIGATING THE RELATIONSHIP BETWEEN CURRENT AND RESISTANCE (OHM'S LAW)

Materials:

a low-voltage d.c. power supply, 0 to 10 volts
five wire-wound resistors, 1, 2, 3, 4 and 5 ohms respectively, all of 10-watt size
an ammeter, 0 to 2A, or similar range
a voltmeter, 0 to 2.5V, or similar range
a mounted knife switch
a circuit board with terminal posts
insulated copper wire

Procedure:

Construct the experimental circuit exactly as shown in Figure 18-11, with the 1-ohm resistor as a load. Keep the power supply turned off, and its voltage control at zero. Check the completed circuit carefully against the schematic diagram. Ask your teacher to inspect the circuit.

Turn on the power supply and close the knife switch. Set the voltage control to exactly 2 volts (on the voltmeter in the circuit) and read the resulting current flow. Record that value. Open the knife switch and exchange the resistor for the 2-ohm device. Close the knife switch, correct the voltage level for 2 volts, and read the resulting current flow on the ammeter. Record that value. Repeat this procedure with the 3-, 4- and 5-ohm resistors, each time applying exactly 2 volts. Then turn off the power supply, dismantle the circuit and return its parts into storage.

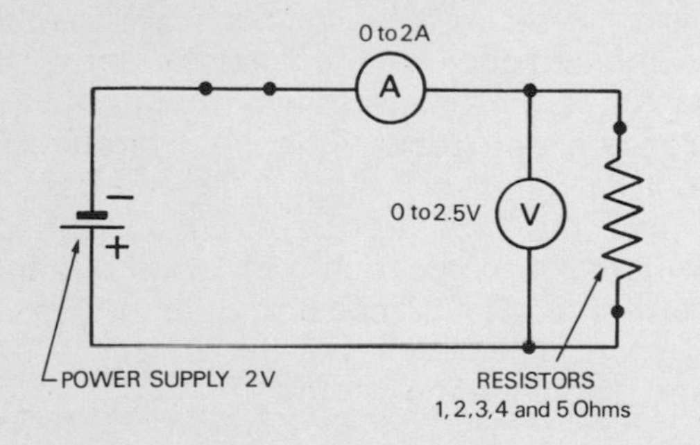

Figure 18-11 Schematic Diagram of Experimental Circuit for Measuring Current vs. Resistance

Assignments:

1. Prepare a neat chart of the current values recorded for each resistance value. Use the following headings: Circuit Resistance, Electron Current, and Applied Voltage.
2. Describe the relationship between circuit resistance and electron current (with a constant applied voltage) which you discovered in this experiment.
3. Explain why an increase in circuit resistance always results in a decrease in current flow.
4. A constant voltage is applied to a 20-ohm resistance and the current flowing through it is 5 amperes. How much current will flow (at the same voltage) if the resistance is changed to a) 10 ohms? b) 40 ohms? c) 5 ohms? d) 100 ohms?

18-11 BUILDING AND ANALYSING A SERIES CIRCUIT

Materials:

a low-voltage d.c. power supply, 0 to 10 volts
two 3-volt lamps in miniature sockets
a knife switch
a suitable wiring board with terminal posts
insulated copper wire
a voltmeter, 0 to 10 volts d.c., with leads

Procedure:

Wire the circuit exactly as shown in the schematic diagram, making all connections neat and tight. Keep the power supply turned off and its voltage control at zero.

Ask your teacher to check the circuit. Then close the knife switch, turn on the power supply, and set its output to 6 volts. Observe and record the brilliancy of the lamps. Next, unscrew lamp 1 and observe what happens to lamp 2. Leaving both lamps aglow, measure accurately the following voltages around the series circuit: the applied voltage (E_A), the voltage drop across lamp 1 (E_1), the voltage drop across lamp 2 (E_2), and the total voltage drop ($E_1 + E_2$).

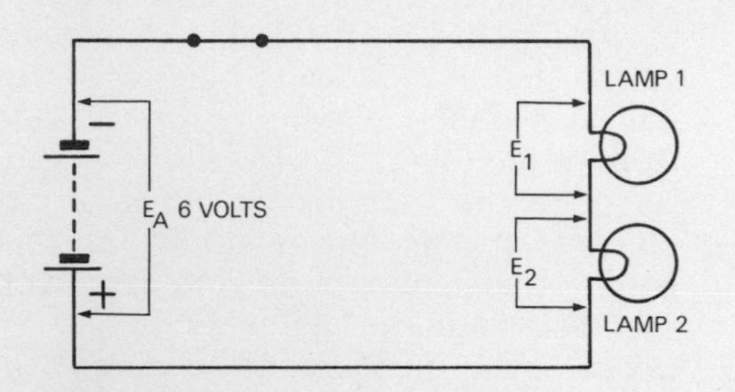

Figure 18-12 Schematic Diagram of Series Circuit

Record all your readings at the time you take them. Then turn the power supply off and dismantle the circuit.

Assignments:

1. Under the heading "Series Circuit," draw a neat schematic diagram of the experimental circuit, labelling all its parts.
2. Make a chart, showing all voltages around the circuit.
3. Write a statement expressing the relationship between the applied voltage and the voltage drops.
4. Express the statement in Assignment 3 as an algebraic equation, using the quantities E_A, E_1 and E_2.
5. Explain what happens in a series circuit if the circuit is broken at any point.

18-12 BUILDING AND ANALYSING A PARALLEL CIRCUIT

Materials:

a low-voltage d.c. power supply, 0 to 10 volts
an ammeter, 0 to 1 ampere d.c., with leads
two 6-volt miniature lamps with sockets
a knife switch
a voltmeter, 0 to 10 volts d.c., with leads
a suitable wiring board with terminal posts
insulated copper wire

Procedure:

Keep the power supply turned off and its voltage control at zero. Wire the parallel circuit, making all connections neat and tight, and ask your teacher to check your work. Insert the lamps into their sockets.

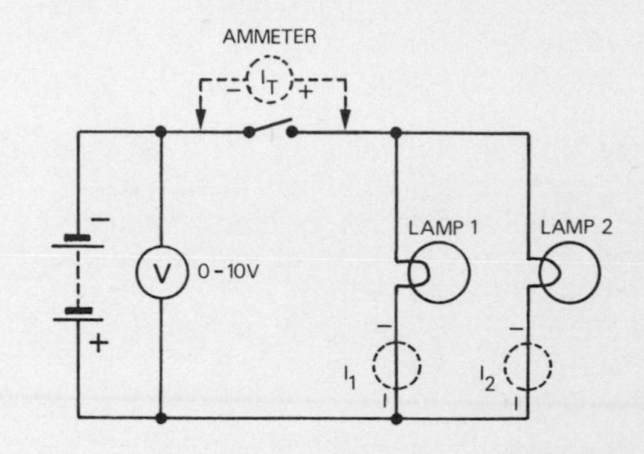

Figure 18-13 Schematic Diagram of Parallel Circuit

Close the knife switch, turn on the power supply, and set its output voltage to 6 volts. Observe the brilliancy of the lamps and record your findings. Unscrew lamp 1 and re-tighten it, observing its effect on lamp 2. Do the same thing with the other lamp and record your observations. Measure the voltage across the following points: the power supply (E_A), lamp 1, and lamp 2. Record these values. Next open the switch and connect the ammeter across the switch terminals, as shown in Figure 18-13. Measure and record the total circuit current, I . Remove the ammeter and connect it into the first branch circuit. Close the switch and measure and record the branch current I_1. Next, measure I_2 in the same fashion. Then turn off the power supply and dismantle the circuit.

Assignments:

1. Under the title "The Parallel Circuit," draw a neat schematic diagram of the experimental circuit. Label all its parts.

2. Compare the applied voltage, E_A, with the voltage across each load, E_1 and E_2.
3. Compare the total circuit current, I, with the individual branch currents, I_1 and I_2. Write an algebraic equation expressing this relationship.
4. Why does each load in a parallel circuit receive all the energy of those electrons flowing through it?
5. How is the electron current in a parallel circuit distributed if all loads are equal?

19

EXPERIMENTS WITH ELECTROMAGNETISM

The experiments described in this chapter should prove to be informative. The extent to which they will work for you depends largely on your approach to them:

> Make certain that you have a clear idea of what it is you want to investigate.
> Obtain all necessary materials before you begin your experiment and check them over carefully. Report defective equipment to your teacher.
> Follow the instructions closely and get involved with what you are doing by concentrating on your work.
> Do the assignments accurately and faithfully. They will help you to remember what you have learned.

19-1 OERSTED'S EXPERIMENT

While doing a similar experiment in 1819, Hans C. Oersted discovered the existence of electromagnetism.

Materials:

a Number 6 dry cell
a 3-foot (1-metre) length of insulated copper wire
a magnetic compass

Procedure:

Connect one end of the 3-foot (1-metre) wire to the negative terminal of the dry cell, leaving the other end free but stripped of insulation. Place the wire on top of the magnetic compass, as shown in Figure 19-1, and observe whether or not the needle reacts. Now, leaving the wire on top of the compass, hold its free end **briefly** against the positive terminal of the cell. Observe the reaction of the compass needle and record your findings.

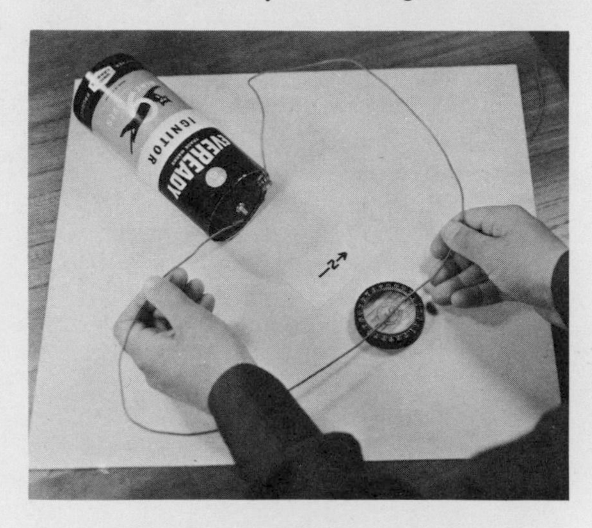

Figure 19-1 Oersted's Experiment

Assignments:

1. Under what conditions does magnetic force appear around the conductor?
2. For how long does the magnetic force around the conductor act?
3. What does Oersted's experiment teach you about free electrons drifting through a conductor?
4. List a few of the thousands of applications of electromagnetism in home and industry.

19-2 INVESTIGATING THE CHARACTERISTICS OF ELECTROMAGNETIC FORCE

Materials:

a 5-inch (12.5-cm.) length of bare AWG 10 wire
a 4 x 4 inch (10 x 10 cm.) piece of cardboard
two Number 6 dry cells, or a charged storage battery
two test leads with alligator clips
six small magnetic compasses
iron filings in a shaker
a jumper wire

Procedure:

Stick the 5-inch (12.5-cm.) piece of AWG 10 wire through the centre of the cardboard. Ask a fellow student to hold it in position, or support it between two books. Using the jumper wire, connect the dry cells in series. Then, connect one test lead from the lower end of the AWG 10 wire to the positive terminal of cell 1 (Figure 19-2). Attach the alligator clip of the other test lead to the top of the AWG 10 wire, but leave its other end free. Now sprinkle a thin, even layer of filings on the cardboard. Hold the bare end of the free test lead against the negative terminal of cell 2 for **1 second** while tapping the cardboard lightly to help the magnetic force align the filings. Repeat this three or four times, until a recognizable pattern becomes visible. (You may have to use fresh cells, or a storage battery.) Draw a neat lines-of-force map based on the filings pattern.

Remove the iron filings and arrange the small compasses around the AWG 10 wire, as shown in Figure 19-3. Again close the circuit for **1 second** and observe the alignment of the compass needles while the current is flowing through the conductor.

Figure 19-2 Investigating the Shape of Flux Lines Around a Conductor

Figure 19-3 Investigating the Direction of Flux Lines Around a Conductor

Reverse the electron flow and repeat the experiment, again observing the alignment of the compass needles. Record your observations, taking care to note the direction of electron flow through the AWG 10 wire.

Assignments:

1. Draw a neat sketch showing the iron-filings pattern around the current-carrying conductor. Explain the shape of the pattern.
2. Make two lines-of-force maps showing the direction of the flux lines when the current flows a) upward b) downward through the wire. (Use tip and tail of arrow to indicate direction of electron flow.)

3. Describe the characteristics of electromagnetic force as observed in the experiment.
4. Make a statement relating the direction of the electromagnetic force to the direction of electron flow.
5. State the left-hand rule for determining the direction of the magnetic force acting around a single conductor.

19-3 DISCOVERING THE INTERACTION BETWEEN MAGNETIC FORCES

The aim of this experiment is to demonstrate what happens when an electromagnetic field interacts with the field of a permanent magnet. This interaction is put to work in hundreds of electrical and electronic devices.

Materials:

a horseshoe magnet
a Number 6 dry cell
a 5-inch (12.5-cm.) length of bare AWG 10 wire
two flexible test leads with alligator clips

Procedure:

Place the horseshoe magnet upright on your desk, as shown in Figure 19-4. Attach the alligator clips of the test leads to the ends of the AWG 10 wire and, holding both wires in one hand above the magnet, form a swing hanging in the centre of the horseshoe magnet's gap. Now, connect one lead to the positive terminal of the cell, and hold the other lead briefly against the negative terminal. Observe and record the reaction of the wire.

Reverse the connections to the cell terminals and repeat the experiment. Again record your findings.

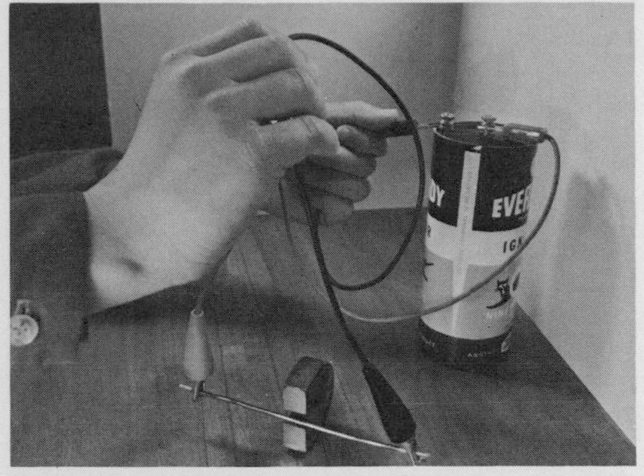

Figure 19-4 Discovering the Interaction Between Magnetic and Electromagnetic Forces

Assignments:

1. Describe briefly how the AWG 10 wire reacted when an electron current was sent through it. Draw simple sketches to illustrate your work.
2. What happened to the AWG 10 wire when the direction of the electron flow through it was reversed?
3. Using the left-hand rule for single conductors and your knowledge of the direction of magnetic force, explain why the wire behaved as it did during the experiment.

19-4 INVESTIGATING MAGNETIC FORCES BETWEEN PARALLEL CONDUCTORS

Materials:

two Number 6 dry cells
two 4-foot (1.3-metre) lengths of magnet wire, AWG 20 or 22
a jumper wire
a 12-inch (30-cm.) wooden ruler, or wooden board of similar size
a hacksaw blade

Procedure:

Cut a ½-inch slot into each end of the ruler, as shown in Figure 19-5. Stretch the 2 magnet wires in parallel along the centre of the ruler, keeping them about 1/32 inch (.75 mm.) apart. Secure the wires by wrapping them around the split ends of the ruler. Now make a rat-tail joint in the wires coming from each end of the ruler, as shown in Figure 19-5. Using the jumper wire, connect the cells in series. Then, connect one joint to the negative terminal of cell 1 and hold the other one **briefly** against the positive terminal of cell 2, observing carefully the reaction of the 2 parallel wires. Write down your findings.

Figure 19-5 Investigating Magnetic Forces Between Parallel Conductors (1)

Now, untwist the 2 joints and make new ones, this time connecting the left end of the first conductor to the right end of the second one (this reverses the current flow through one of the wires). Join the remaining ends also and connect them to the negative terminal of cell 1. Touch the other joint **briefly** against the positive terminal of cell 2 and watch the reaction of the parallel wires (it will be a feeble reaction). Record your observations.

Figure 19-6 Investigating Magnetic Forces Between Parallel Conductors (2)

Assignments:

1. In the first part of the experiment, both wires carried electron currents flowing in the same direction. Explain the result.
2. In the second part, the electron currents flowed in opposite directions through the parallel wires. What was the result?
3. In the experiment, the forces tugging at the wires were barely detectable. Why?
4. Summarize the results of the experiment in a single statement.

19-5 EXPLORING THE MAGNETIC FORCE OF A COIL

Materials:

two Number 6 dry cells, or a storage battery
a jumper wire
a 2-foot (60-cm.) length of bare AWG 14 wire
a pair of test leads with alligator clips
a piece of cardboard, 4½ x 4½ inches (11.2 x 11.2 cm.)
ten small magnetic compasses
iron filings in a shaker
a D size dry cell, or a cylindrical object of similar size

Procedure:

Using the D cell as a core, wind the AWG 14 wire into 5 tight turns. Punch 2 rows of holes into the cardboard piece, 1¼ inches (3.1 cm.) apart and spaced ⅛ of an inch (3 mm.). Thread the coil into the cardboard holes and bend its ends outward, as shown in Figure 19-7.

Figure 19-7 Exploring the Magnetic Force of a Current-Carrying Coil

Use the jumper wire to connect the cells in series. Connect one test lead from the coil to the negative terminal of cell 1, and clip the second lead to the other end of the coil without connecting it to the cells.

Sprinkle a thin, even layer of iron filings on the cardboard and **briefly** hold the second test lead on the positive terminal of cell 2. Tap the cardboard lightly. Repeat this 2 or 3 times until the filings form a recognizable pattern. Record your observations and draw a sketch of the iron-filings pattern.

Figure 19-8 Showing the Direction of a Coil's Magnetic Force

Remove the iron filings and replace them with the 10 small magnetic compasses, as shown in Figure 19-8. Close the circuit for one second and observe the reaction of the compass needles. Make a sketch of the set-up, indicating the direction of electron flow in the coil, the alignment of the needles, and the N- and S-pole of the coil's magnetic field.

Reverse the connection of the test leads to the battery and repeat the experiment.

Assignments:

1. Make neat copies of the sketches containing the iron-filings pattern and the data obtained in the second part of the experiment.
2. Describe the similarities between the magnetic field of a current-carrying coil and that of a permanent magnet.
3. Explain the relationship which exists between the direction of the coil's magnetic field and the direction of the electron flow.
4. State the left-hand rule for current-carrying coils and electromagnets.

19-6 INVESTIGATING THE PROPERTIES OF ELECTROMAGNETS

Materials:

two Number 6 dry cells
one 4-foot (1.3-metre) length of AWG 22 magnet wire
two metal rods, ¼ inch (6 mm.) in diameter and 4 inches (10 cm.) long, one of copper and the other of iron
a nickel coin
a large magnetic compass
a piece of chalk (or dowel)
some plastic tape
a jumper wire

Procedure:

Scrape 1 inch (2.5 cm.) of insulation from both ends of the magnet wire. (The **length** of this wire **must not be changed** during the experiment.) Wind 10 turns on the piece of chalk and secure the coil with tape. Place the magnetic compass on your desk and wait until the needle points

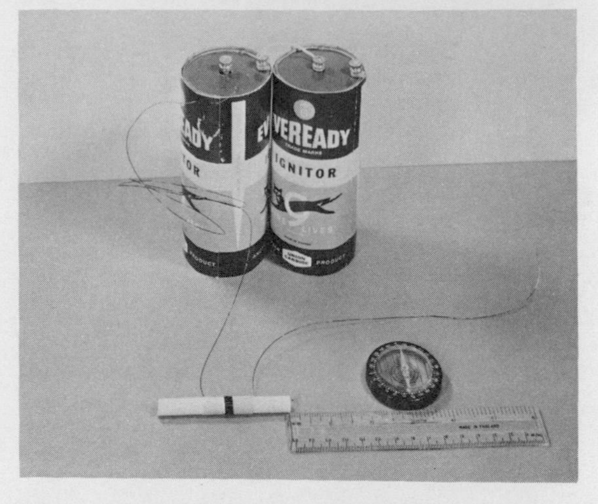

Figure 19-9 Testing the Magnetic Force of an Electromagnet

north. Put the coil on your desk, exactly 3 inches (7.5 cm.) away from the compass's N-pole and at right angles to it (Figure 19-9). Connect one lead to the positive terminal of a cell, and hold the other **briefly** against the negative terminal. Observe and record the number of degrees which the compass needle is deflected by the electromagnetic force.

Using the jumper wire, connect the 2 cells in series. Now, holding the coil in the same position as before [3 inches (7.5 cm.) from the compass], connect it briefly to the 3-volt supply, thus **doubling** the current flow. Again observe and record the deflection of the compass needle.

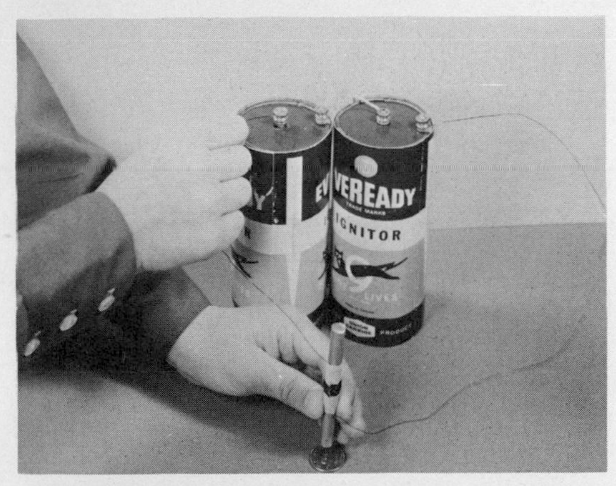

Figure 19-10 Testing Various Cores for an Electromagnet

Remove the tape from one end of the coil and add 10 more turns, winding them on top of the first layer, and secure them with tape. Place the coil in its previous position [exactly 3 inches (7.5 cm.) from the compass] and repeat the experiment. Record your observations, including the deflection of the needle.

For the third part of the experiment, wind 30 closely spaced turns on the chalk and secure the coil with tape. Connect one lead to the negative terminal of the 3-volt battery. Hold the coil ⅛ inch (3 mm.) above the coin (Figure 19-10) and close the circuit for one second. Record your observations.

Unwind the coil and rewind it on the copper rod. Repeat the experiment.

Again unwind the coil, rewind it on the iron rod, and repeat the experiment. Record all your observations.

Assignments:

1. Under the heading "Factors Affecting the Magnetic Force of a Coil" explain:
   a) the relationship between the number of turns and the strength of the electromagnetic force
   b) the relationship between the electron current flowing through a coil and the strength of the electromagnetic force
   c) the relationship between the type of core material and the strength of the magnetic force
2. Explain carefully the meaning of the following statement: the strength of the magnetic force of a coil is directly proportional to the rate of current flow, the number of its turns, and the permeability of its core.
3. Why do the heavy lifting magnets used in many scrap-iron yards have soft-iron cores?
4. Define the term "permeability" and explain its significance in working with electromagnets.

20
PROJECTS

The nine projects in this chapter are not intended to cover the full range of possible construction activities available to the beginning student of electricity. Their main purpose is to familiarize you with basic circuit wiring and component layout techniques.

With the exception of Project 1, no layout diagrams are provided, because each builder will use whatever wiring boards and components are available to him.

Reliability, accuracy, neatness and effectiveness should be your goals in building these projects. Speed of construction is of secondary importance only.

Here are some technical hints:

> A good first step in building a project is to study the outline and to make a layout diagram of the circuit so that you can estimate your requirements.
>
> Gather all the materials for the project before you actually begin its construction.
>
> When wrapping wire around a screw terminal, wrap it clockwise around the screw. In this way, the wire will not be squeezed out when you tighten the screw.
>
> Before you turn on the power, check the completed project carefully against the schematic diagram. Then ask your teacher to approve your work. This technique will prevent many problems.

20-1 SIMPLE BELL CIRCUIT

Materials:

a Number 6 dry cell
a wiring board
a push-button switch
a bell
insulated copper wire, AWG 18, white and black
a diagonal cutter, a knife, and a screwdriver

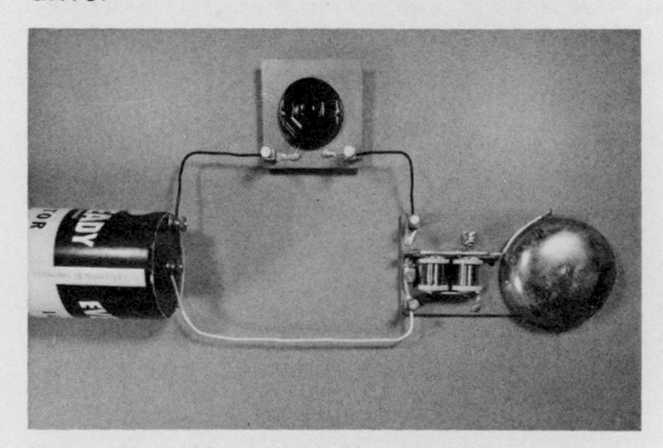

Figure 20-1 Simple Bell Circuit

Procedure:

Using the schematic diagram as a guide, prepare a layout similar to the arrangement shown in Figure 20-1. The components may be attached to the wiring board or simply laid on it.

The practicing electrician uses the colour of a wire's insulation to indicate its function: black — feeder wire; also black — control wire; white — return wire. Use these colours in this project.

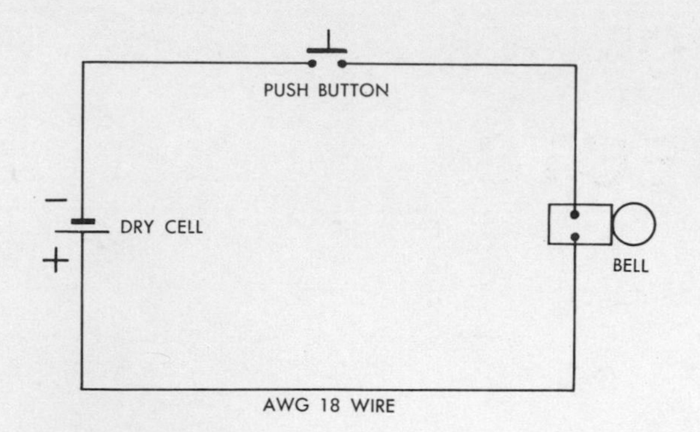

Figure 20-2 Schematic Diagram of Simple Bell Circuit

Assignments:

1. Draw a neat schematic diagram of the simple bell circuit. Label all components and indicate the feeder, control and return wires.
2. Identify and explain the function of the four main components of the circuit.
3. Why is a bell transformer used in place of the dry cell in an actual home installation?

20-2 BELL CIRCUIT WITH CONTROLS IN SERIES

Materials:

a bell transformer (or a low-voltage power source, 6 to 8V)
a bell
two push-button switches
a wiring board
insulated copper wire, AWG 18, white and black
a diagonal cutter, a knife, and a screwdriver

Procedure:

Examine carefully the schematic diagram, Figure 20-3, and make a layout drawing of the actual circuit as you plan to construct it. (The two push-button switches need not be close to each other.)

Build the circuit, using the schematic diagram as a guide. Do not plug in the bell transformer (or low-voltage power supply) until the circuit has been checked both by yourself and by your teacher.

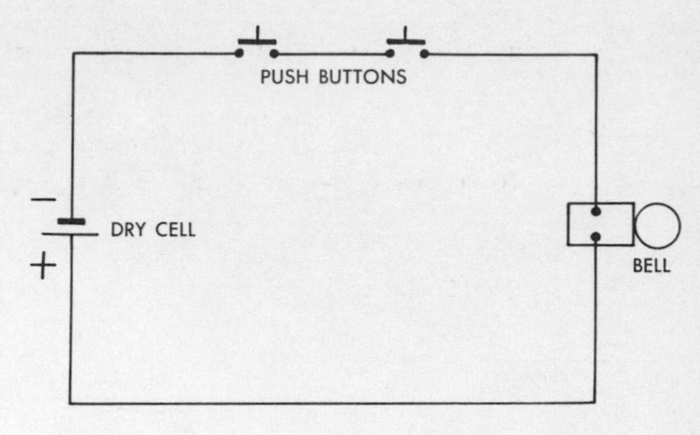

Figure 20-3 Bell Circuit with Controls in Series

Assignments:

1. Draw a neat schematic diagram of the circuit. Label all parts and indicate the function of each wire.
2. Explain why this kind of circuit would be quite impractical for a home signaling system.
3. In industry, switches in series are called a "coincidence circuit" or an "AND gate." Can you explain these names?
4. Name several possible applications of the coincidence circuit in industrial operations.

20-3 BELL CIRCUIT WITH CONTROLS IN PARALLEL

Materials:

a bell transformer (or low-voltage power supply, 6 to 8V)
a bell or buzzer
two push-button switches
wiring board
AWG 18 copper wire, white and black
a diagonal cutter, a knife, and a screwdriver

Procedure:

Study the schematic diagram, Figure 20-4, and prepare the layout of the components. (The 2 push-button switches should be located close to each other.)

Build the circuit, using the schematic diagram as a guide. Do not turn on the power until you have checked the circuit's wiring and have asked your teacher to approve it.

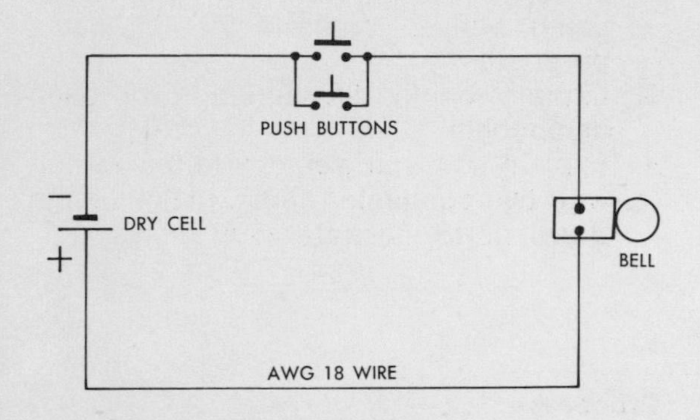

Figure 20-4 Bell Circuit with Controls in Parallel

Assignments:

1. Under the heading "Bell Circuit with Controls in Parallel," draw a neat schematic diagram of the circuit and label all its parts, including the conductor runs.
2. Briefly explain the function of each circuit component, including the wire runs.
3. Explain why the bell can be controlled by either push-button, or by both of them.
4. Where could such a circuit be used? Mention two applications and describe each one briefly.
5. In industrial control circuits, an arrangement of two or more switches in parallel is called an "OR gate." Can you explain this name?

20-4 FRONTDOOR-BACKDOOR SIGNAL CIRCUIT

Materials:

a bell transformer (6 to 8 VAC)
two push-button switches
a buzzabell or a two-gong chime
AWG 18 copper wire, black and white
a circuit board
a diagonal cutter, a knife, and a screwdriver

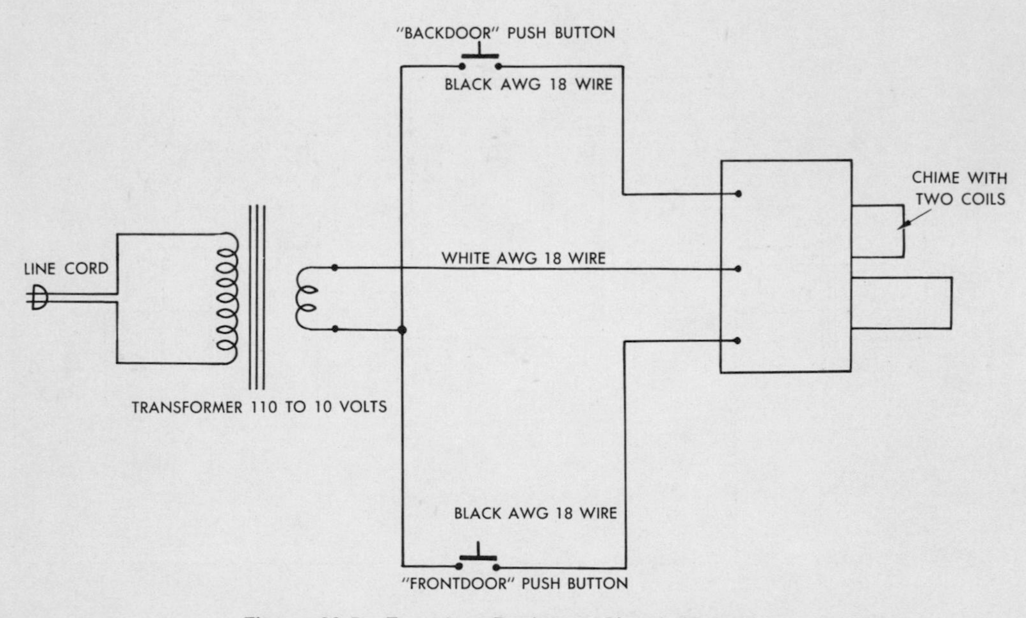

Figure 20-5 Frontdoor-Backdoor Signal Circuit

Procedure:

Plan the layout of the circuit components based on the schematic diagram, Figure 20-5. Locate the signal device on the right side of the circuit board, the bell transformer on the left, and the push-button switches approximately as indicated on the schematic diagram.

Check the completed circuit carefully against the schematic diagram and ask your teacher to inspect it before you turn on the power.

Assignments:

1. Under the heading "Frontdoor-Backdoor Signal Circuit," draw a neat schematic diagram of the circuit. Label all components, including the wire runs.
2. State the main advantages of this circuit when compared with a circuit using parallel controls for the same purpose.
3. Explain briefly the function of each component.
4. Explain how you would wire this circuit with two-conductor cable. Make simple drawings to illustrate your answer.

20-5 ANNUNCIATOR CIRCUIT

Materials:

a four-point annunciator
a bell transformer, 12 to 18 volts
four push-button switches
AWG 18 wire, black and white
a circuit board
a diagonal cutter, a knife, and a screwdriver

Procedure:

Study the schematic diagram, Figure 20-6, before you plan the layout of the circuit components. Then prepare a layout plan and discuss it with your teacher.

Wire the circuit with great care, making certain that all connections are reliable. Compare the completed project with the layout and the circuit diagram to discover possible wiring errors. Then ask your teacher to check your work before you turn on the power.

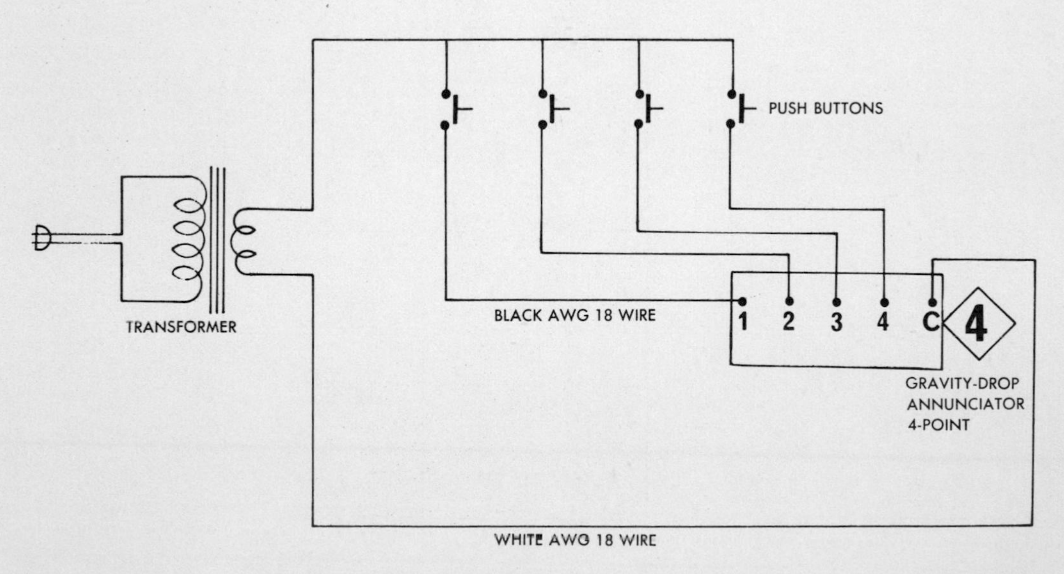

Figure 20-6 Four-Point Annunciator Circuit

Assignments:

1. Draw a neat schematic diagram of the annunciator circuit. Label all components, including the wire runs.
2. Describe how the annunciator works and why it must be reset each time a number has dropped.
3. List several typical applications for annunciators.
4. Draw a neat wiring diagram showing how you would wire this circuit with two-conductor cable.

20-6 TWO-WAY MORSE CODE SIGNAL CIRCUIT

Materials:

two bell transformers, 6 to 8 VAC
two 6-volt miniature bulbs with sockets
two push-button switches
two buzzers
two wiring boards
3-conductor cable (3 x AWG 18)
AWG 18 wire, black and white
two terminal boards with three terminal screws each (or binding posts)
a diagonal cutter, a knife, and a screwdriver

Procedure:

Examine the circuit diagram, Figure 20-7, with great care and plan the layout of the components for both circuits. Discuss this plan with your teacher.

Wire each signal station on a separate circuit board. Inspect the completed circuits for possible faults by comparing them with the schematic diagram. Ask your teacher to approve your work before you plug in the bell transformers.

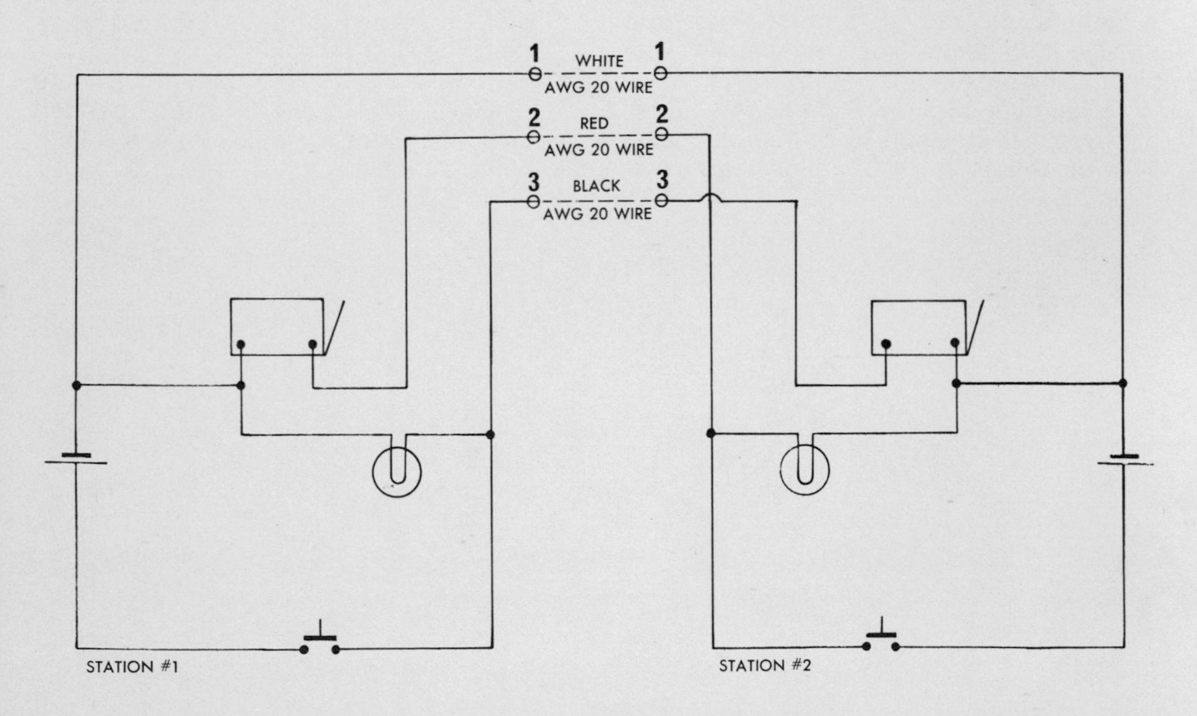

Figure 20-7 Two-Way Morse Code Signal Circuit

Assignments:

1. Make a neat, fully labelled schematic diagram of the two-way morse code circuit.
2. Explain how the circuit works, using the schematic diagram as a reference.
3. Why would the voltage of the two transformers have to be increased if the two stations were located very far apart?

20-7 SIMPLE RELAY CIRCUIT

Materials:

a silicon solar cell, mounted
a low-voltage relay, with normally open contacts
a Number 6 dry cell (or a D cell in a holder)
a buzzer
a flashlight
a wiring board
AWG 18 wire, white and black
a diagonal cutter, a knife, and a screwdriver

Procedure:

Draw a layout plan of the circuit, using the schematic diagram as a guide. The relay must be a special type designed to work with a silicon solar cell (0.45 VDC).

Build the circuit with great care. When it is completed, test it by shining the flashlight on the solar cell. Turning the flow of light energy on and off should also turn the buzzer on and off.

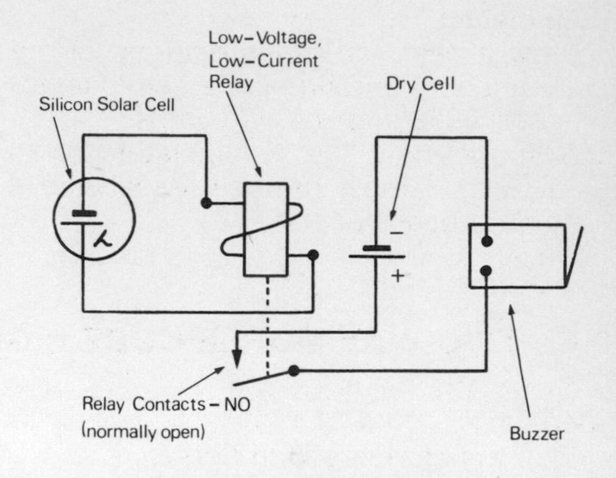

Figure 20-8 Simple Relay Circuit

Assignments:

1. Draw a neat schematic diagram of the circuit and label all components, including the wire runs.
2. Describe carefully how the circuit works.
3. State a few possible uses of such a circuit (or for two of them, located some distance apart).
4. How would you make this circuit more sensitive, or capable of being operated over a greater distance from the light source?

21 THE LANGUAGE AND SYMBOLS OF ELECTRICITY

In the four hundred years since William Gilbert began his research into the nature of electricity and magnetism, a huge specialized vocabulary has been built up in this field. This "language of electricity" includes a system of formulae and schematic symbols which are designed to assist the practicing technician in his work.

The terms and symbols listed in this chapter have been carefully selected for the beginning student of electricity. The definitions are simple and direct. The formulae and schematic symbols reflect current international standards.

21-1 THE SCHEMATIC CIRCUIT DIAGRAM

A schematic diagram is a greatly simplified representation of an electric circuit. It shows only those parts which are essential to the working of the circuit. Mounting devices, clamps, insulators, sockets, etc. are not shown at all, and every circuit component is represented by a greatly simplified **schematic symbol.**

To the practicing electrician, the schematic circuit diagram is the single most important aid in wiring circuits and in building electrical devices. A reliable knowledge of the most common schematic symbols therefore is required information for anyone working with electric circuits.

In your notebook — or for any other official purpose — schematic circuit diagrams should always be drawn with the aid of a straight-edge and a compass, or by using the special plastic templates available for this purpose.

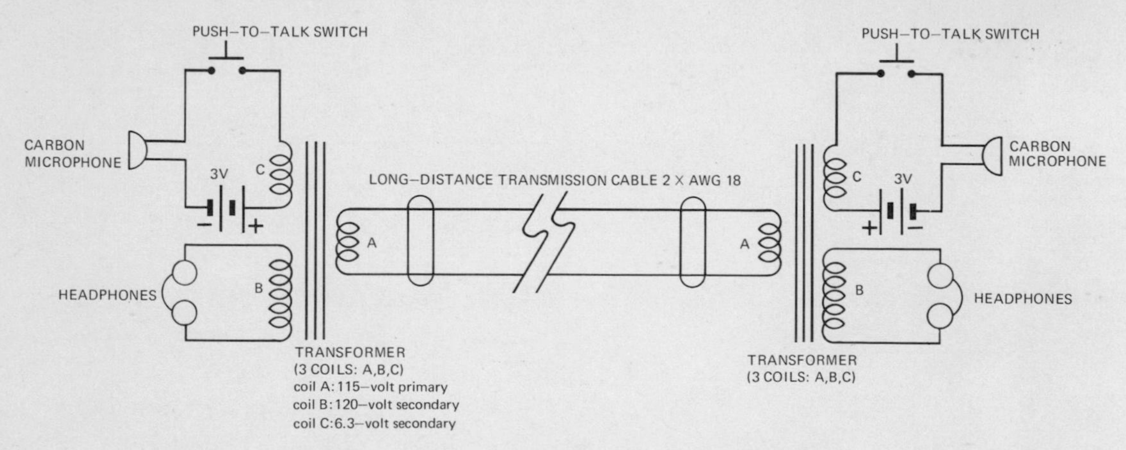

Figure 21-1 Schematic Diagram of an Electric Circuit

21-2 COMMON SCHEMATIC SYMBOLS

The schematic symbols included in this list are commonly used in North America; they do not reflect international standards. Many large corporations and some professional groups use a system of "in-house" symbols that are different from those shown here. However, there is evidence of a trend toward universal standardization of schematic symbols, and such fine points as indicating a junction of conductors by means of a heavy dot are now gaining wide-spread acceptance.

CONDUCTORS

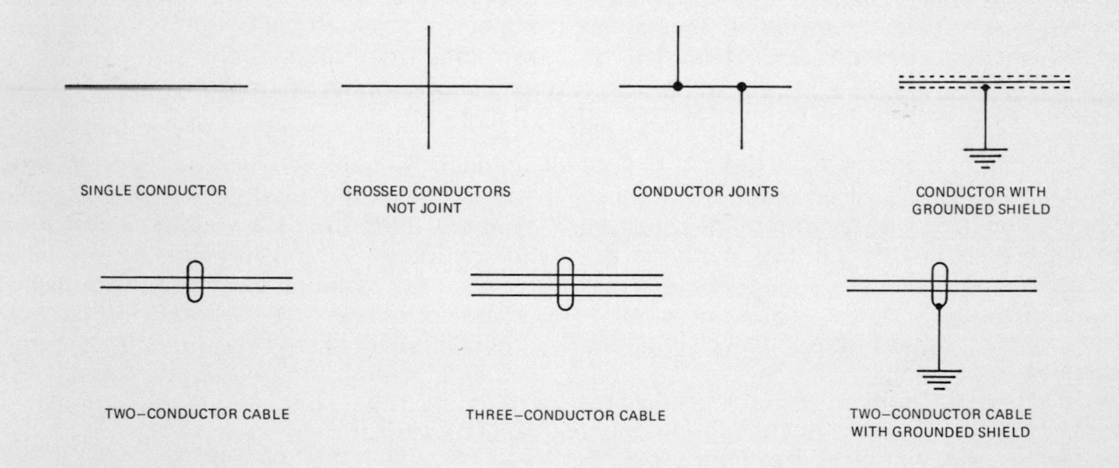

CONNECTORS

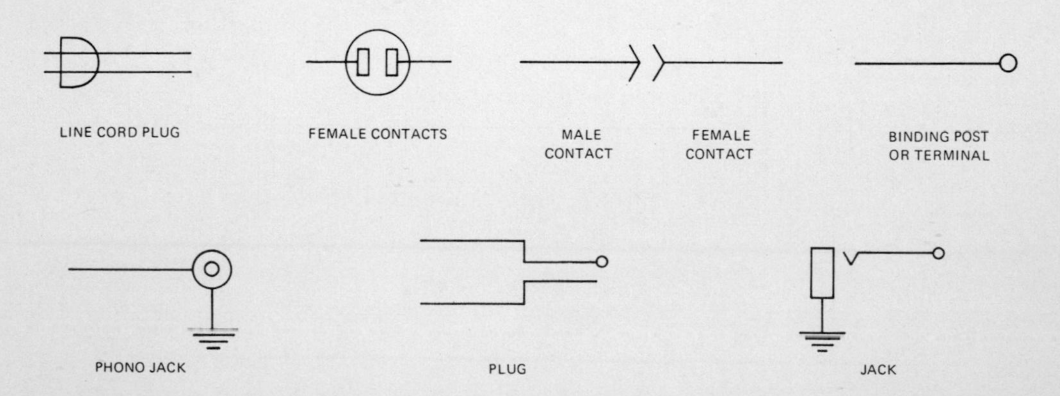

ENERGY SOURCES (CONVERTERS)

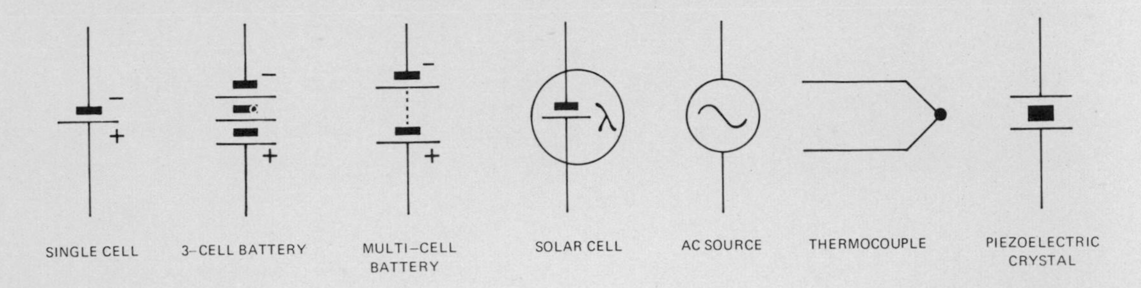

CONTROLS

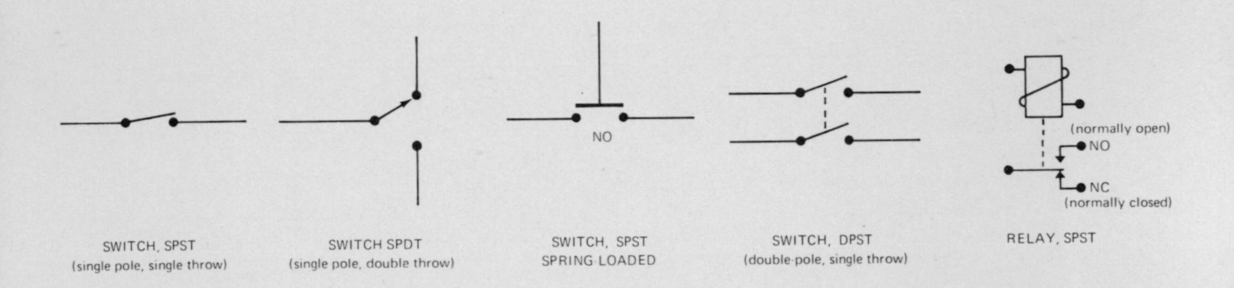

RESISTORS

LAMPS

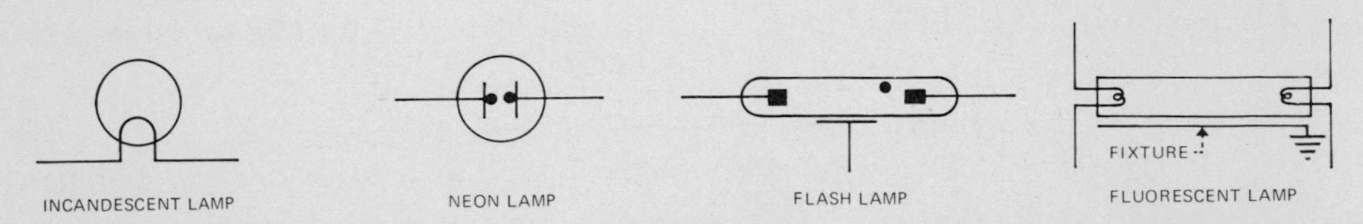

TRANSFORMERS

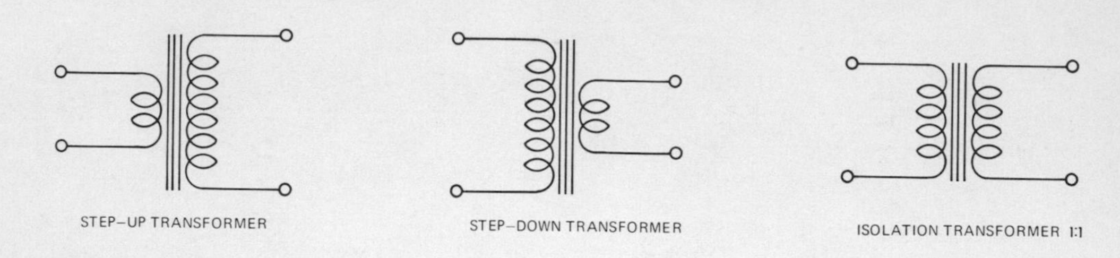

TRANSDUCERS

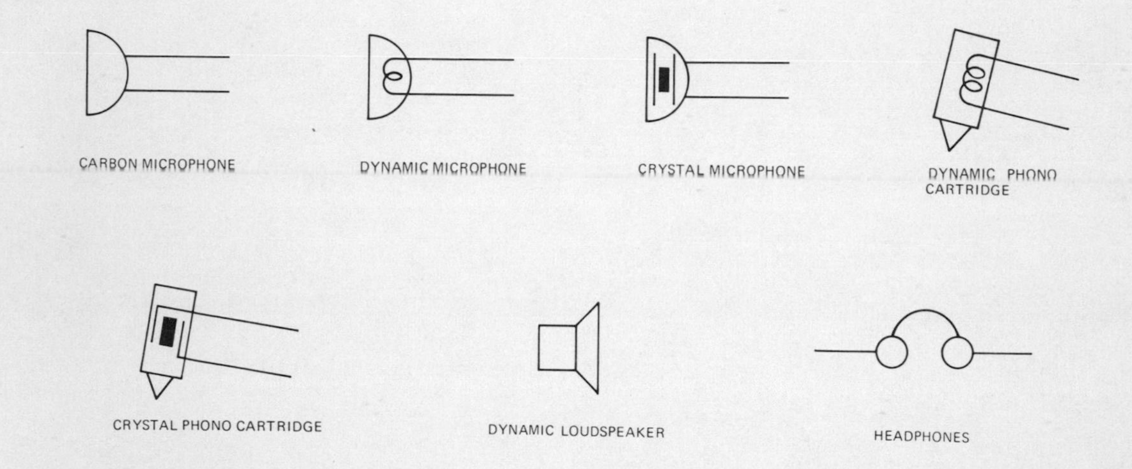

CIRCUIT PROTECTION DEVICES

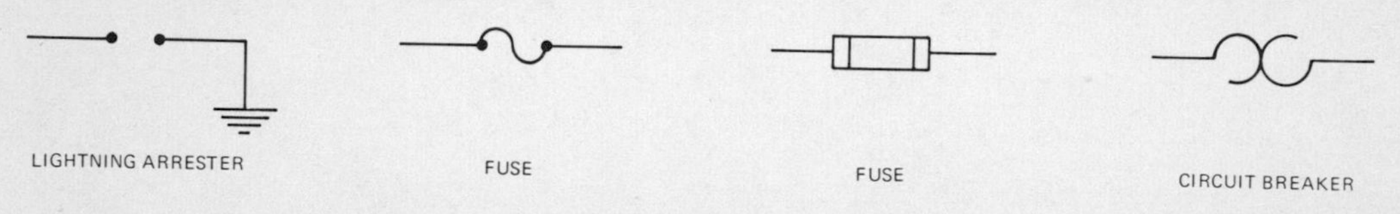

Figure 21-2 Common Schematic Symbols

21-3 ELECTRICAL TERMS

The following is a short dictionary of important terms for the beginning student of electricity. If a term listed in this dictionary occurs within a definition, it is **italicized.** This will help you to find related terms as quickly as possible.

Use this dictionary often to familiarize yourself with its contents. Its terms are of basic importance to your study of electricity.

a.c. Abbreviation for *alternating current.*
air core coil A copper wire *coil* wound on a plastic core or *bobbin.*
alnico magnet A powerful *permanent magnet* which is composed of aluminum, nickel and cobalt. It can be cast in any shape or size and will retain its *magnetism* indefinitely.
alternating current Abbreviated *a.c.* An *electric current* which continuously varies its intensity from zero to a maximum value back to zero, and which reverses its direction of flow at regular intervals. Each complete back-and-forth flow is called a *cycle,* and the number of *cycles per second* is called the *frequency* of the alternating current.
alternator Rotary *a.c.* generator used in hydroelectric power plants to generate a.c. for domestic and industrial use.
amber The petrified sap of fir trees. Its Greek name is elektron. According to legend, the effects of *static electricity* were first observed when pieces of amber were rubbed with fur. The modern terms *electron* and *electricity* are derived from the Greek term for amber.
American wire gauge Abbreviated *AWG.* A system of standardized wire sizes used for electrical *conductors.* Sizes range from AWG 0000 (0.46 inches in diameter) to AWG 50 (0.001 inches in diameter). The most common size used in house wiring is AWG 14 (0.0641 inches in diameter).
ammeter Short form for *ampere* meter. An instrument for measuring the amount of *current* flowing in a conductor.
ampere Formula symbol: **A.** *Metric (S.I.)* unit of *current* flow. One ampere equals a flow of 6.247 billion billion electrons per second, which is one *coulomb* per second.
ampere-hour Formula symbol: **a.h.** *Metric (S.I.)* unit of measurement for cell *capacity.* A dry cell or battery has a capacity of one ampere-hour if it can supply a current flow of one ampere for a period of one hour. One a.h. equals 3,600 *coulombs* of *charge.*
ampere-turn Formula symbol: **a.t.** The amount of *magnetomotive force* produced by an *electron current* of one ampere flowing through one turn of a *coil.*
AND gate Computer term for two or more switches in series; also called coincidence circuit.
annunciator *Electromechanical signal device* using number tabs to indicate where a signal originated.
armature The movable part of a *relay,* bell or *buzzer;* operates the *interrupter contacts* or the switch contacts of a relay.
artificial magnet *Permanent magnet* made of *alnico* alloy.
atom Organized system of *elementary particles. Protons* and *neutrons* form the small, dense and heavy *nucleus* which contains nearly the entire mass of the atom. Orbiting *electrons* form the *electron cloud* which cannot be compressed to smaller size.
atomic magnets The *atoms* of iron, nickel and cobalt are tiny *permanent magnets* with an *N-pole* and *S-pole.* These atomic magnets can be swivelled into alignment by an external magnetic force.
atomic structure The internal structure common to all *atoms,* consisting of a small, dense and heavy *nucleus* surrounded by an *electron cloud.*
AWG Abbreviation for *American Wire Gauge.*
battery A *d.c. voltage* source consisting of one or more energy converter cells such as *primary, secondary* or *solar cells.*
bell transformer Small *transformer* of special design, with built-in *overload* and *short-circuit* protection. Available in two *output voltage* ranges, from 6 to 10 volts a.c. and from 12 to 18 volts a.c.
bi-metallic strip A metal strip consisting of two unlike metals welded together. The two metals expand or contract at different rates when the strip is heated or cooled, causing the metals to bend up or down. This bending action can be used to operate *switches* and other *control* devices.
binding post A screw terminal for electrical connections.

bobbin Plastic tube or core for holding the *coil* of an *electromagnet.*

branch circuit Part of a *parallel circuit;* can operate independently from the other branch circuits.

break Any accidental interruption in the metallic pathway of an *electric circuit.*

buzzabell *Signal device* combining a bell and a *buzzer* on a common mounting plate.

buzzer *Signal device* consisting of a vibrating *armature* with *interrupter contacts* and two small *electromagnets.* When operated, the continual attraction and release of the spring-loaded armature creates a buzzing sound.

cable Two or more insulated and flexible *conductors* contained in a common insulating sleeve.

Canadian Electrical Code Abbreviated **CEC.** A body of rules and regulations governing the design, construction and installation of electrical circuits and components in Canada. The CEC is incorporated in the various provincial electrical codes.

Canadian Standards Association Abbreviated **CSA.** A government-sponsored testing and approval agency for electrical components and equipment used in Canada. All devices and equipment installed in electrical systems must bear the stamp of approval issued by the CSA.

capacity The total electric *charge* contained in the chemicals of a *dry cell* or *storage battery.* Cell capacity is measured in *ampere-hours,* a.h.

cartridge fuse Tube-shaped *fuse* for the protection of electric circuits, available in many sizes and with both disposable and renewable *fuse links.*

CEC Abbreviation for *Canadian Electrical Code.*

cell A single unit of a *battery.* Also, an *energy converter* such as a *primary, secondary* or *solar cell.*

cell capacity See *capacity.*

charge Formula symbol: **Q.** Lack or surplus of *electrons* on a body. Unit of measurement: the *coulomb.* Also, the unique property of electrons and *protons* which enables them to exert an *electric* force in the surrounding space.

circuit Closed-loop metallic pathway for *free electrons,* from the *negative* to the *positive terminal* of a *power supply.*

circuit breaker A *circuit protection device* which opens the circuit automatically when an *overload* or *short circuit* occurs.

circuit diagram A drawing which represents all parts of an *electric circuit* in *pictorial* or *schematic* (symbolic) form.

circuit fault Any defect in an *electric circuit* or *component.*

circuit protection device A *fuse, circuit breaker, lightning arrester* or *lightning rod* which protects a circuit from damage caused by *overload* or *short-circuit* currents.

closed circuit An *electric circuit* which forms a complete, uninterrupted pathway for *free electrons.*

code Short form for *Electrical Safety Code.*

coil One or more turns of bare or insulated copper wire which, when carrying an *electron current,* produce an *electromagnetic field* having an *N-pole* and *S-pole* exactly like a *bar magnet.*

compass See *magnetic compass.*

complex circuit Any *electric circuit* which combines *series* and *parallel* elements in various interconnections.

component Any electrical device that can be connected to an *electric circuit* or made a part of it, such as bells, *buzzers, fuses, switches, resistors,* etc.

conductivity The ability of a metal to permit the flow of *free electrons* through its internal structure; the opposite of *resistance.*

conductor Any material containing *free electrons,* but usually assigned only to silver, copper, gold and aluminum wires. Also, any wire or *cable* which can carry an *electron current.*

contact That part of a *switch* or *relay* which actually makes or breaks the circuit. Contacts are made of a special alloy which is resistant to *pitting* and corrosion.

continuity tester A simple *battery-buz-*

zer device which is used to send a small testing current through a *circuit* to detect possible breaks in the metallic pathway.

control A *component* or device which starts, stops or controls the *current* flow in an *electric circuit.*

control circuit In a *relay* circuit, the low-current path which is used to energize the relay coil.

control wire The *conductor* leading from the *control* to the *load* in an *electric circuit.*

conventional current Term used for the direction of current flow originally assumed by Benjamin Franklin. Conventional current is thought of as running opposite to the actual *electron flow.*

core The *magnetic* soft-iron core of a *transformer, relay, solenoid* or *electromagnet.*

coulomb Formula symbol: **C.** *Metric (S.I.)* unit of electric *charge,* equal to the combined charge of 6.247 billion billion *electrons.* Named in honour of the French physicist C.A. de Coulomb.

crimp-on connector A special solderless *wiring connector* which is attached to a *conductor* by means of a *crimping tool.*

crimping tool A pair of special pliers that can exert great force on the sleeve of a *crimp-on connector* to attach the connector permanently to a *conductor.*

CSA Abbreviation for *Canadian Standards Association.*

CSA approval A special stamp or sticker issued by the *Canadian Standards Association* for attachment to electrical products which have been approved by that agency.

current Formula symbol: **I.** Short form for *electric current* or *electron current,* the rate of *charge* flow through a *conductor* or through a conducting solution (electrolyte). Unit of measurement: the *ampere.*

cycle One complete back-and-forth flow of *electrons* in an *a.c.* circuit, from zero to maximum current back to zero in one direction, and from zero to maximum current back to zero in the opposite direction in a rising and falling, wave-like motion.

cycles per second Formula symbol: **CPS.** English unit of measurement for the *frequency* or repetition rate of an *alternating current.* Its *metric (S.I.)* equivalent is the *hertz,* which is equal to one *cycle per second.*

d.c. — direct current *Electron current* which always flows in the same direction through an *electric circuit,* propelled by an *emf* that does not change its *polarity.*

d.c. generator A rotating converter of mechanical to electrical energy whose output is *direct current.*

d.c. power supply A device which changes the 115-volt *alternating current* from an outlet to a controllable d.c. voltage, generally below 30 volts.

dead circuit An *electric circuit* from which the *voltage* source has been removed, either intentionally or through a *break* in the *circuit.*

dead short A *short-circuit* path of extremely low *resistance.*

depolarizer Chemical substance (manganese dioxide) used in carbon-zinc *dry cells* to prevent the accumulation of hydrogen gas around the carbon rod.

dielectric Any insulating material such as glass, ceramics, plastics, etc., which does not conduct *electricity.*

discharge The flow of *electricity* through a gas, usually accompanied by a visible glow or an arc. Also, to remove or use the *charge* stored in a *secondary cell.*

door chime *Electromagnetic signal device* which produces a gong sound when operated.

doping Introducing an impurity into a pure *semiconductor* material to change the semiconductor to a P-type or N-type.

dry cell A primary or *voltaic cell* whose *electrolyte* is in paste form to prevent it from spilling during use.

dynamic electricity *Electric charges* in motion; usually, a flow of *electrons* through a *conductor,* carrying *energy* from the *supply* to the *load.*

eddy current Undesirable circulating *currents* induced in the soft-iron core of a *transformer* by the changing *magnetizing force;* generates heat.

eddy current loss The *energy loss* in a *transformer core* due to circulating *eddy currents;* it is dissipated in the form of heat.

Electrical Safety Code A body of rules and regulations governing the design and construction of electrical circuits and equipment in Canada, and amended by the various provincial authorities; it is published by the *CSA.*

electric charge Alternate name for *charge.*

electric circuit Alternate name for *circuit.*

electric current Alternate name for *conventional current* or *electron current.* The term electric current is used when the actual direction of *electron flow* is not important to the user.

electric energy Formula symbol: **W.** The *energy* carried by *free electrons* from a *source* to a *load.* Also, the *potential energy* of a stationary *electric charge.*

electric force The invisible force which an electrically charged body exerts in the surrounding space.

electricity General name used for all forms of *electric energy*. Also, the name of the field of study dealing with *electric energy.*

electric power Formula symbol: **P.** *Electric energy* at work.

electrolyte The conducting solution used in *primary* and *secondary cells.*

electromagnet A current-carrying *coil,* usually with a soft-iron *core* to concentrate and increase its *magnetic force*. The *magnetomotive force* of the *electron current* aligns the *magnetic atoms* in the core and sets up a powerful *field of* magnetic *force* in the surrounding space.

electromagnetic induction The act of propelling *free electrons* into motion through a *conductor* by means of a moving magnetic force cutting through the conductor.

electromagnetism The invisible magnetic force generated when *free electrons* flow through a *conductor.* More generally, the *magnetism* generated by *free electrons* in motion.

electromechanical Using *electricity* to do mechanical work, such as in electric motors or in *solenoids.*

electromotive force Formula symbol: **EMF** or **E.** Alternate name for *voltage* or potential difference. The force which propels *free electrons* through a *circuit.* Unit of measurement: the *volt.*

electron Smallest and lightest of the *elementary particles.* Each electron has a permanent *negative charge* which causes a field of *electric force* to exist in the space surrounding the particle.

electron cloud The outer part of an *atom;* it is composed of *electrons* orbiting the central *nucleus.* The electron cloud cannot be compressed to smaller size, but electrons can be removed from it or forced into it.

electron current The flow of *free electrons* through a *conductor,* propelled by an *electromotive force.* See also *current.*

electron flow Alternate name for *electron current. Free electrons* always flow from the *negative* terminal of a power supply through the *circuit* to the *positive terminal* of the *supply.*

electrostatic Pertaining to stationary *electric charges* which set up a field of *electric force* in the surrounding space.

elementary particle A term used for *electrons, protons* and *neutrons,* and for several sub-nuclear particles.

emf Abbreviation for *electromotive force.*

endpoint voltage The *voltage* level which indicates that a *dry cell* must be replaced. Its value depends on the use of the *cell.*

energy Formula symbol: **W.** The ability to do work. Unit of measurement: the *joule* or *watt-second.*

energy converter Any device which converts *energy* from one form to another.

feeder wire Name given to the *conduc-*

tor which runs from one *terminal* of a *power supply* to the *control* device (switch) in an *electric circuit.* In house wiring, a wire with black insulation is chosen for this purpose.
field of force A region in space in which *electric* or *magnetic force* acts or exists, usually in the immediate vicinity of a charged body or of a *magnet* or *electromagnet.*
filament The finely coiled tungsten wire in an *incandescent lamp* which when heated gives off a brilliant white light.
flashover A sudden electric *discharge* around or over the surface of an *insulator* due to excessively high *voltage.*
flux The invisible lines of *magnetic* or *electric force* existing in the space around a *magnet* or a *charged body.* Also, the rosin used in soldering to prevent the forming of an oxide film on the surfaces to be soldered.
flux lines Alternate name for the *lines of magnetic force* existing in the space around a *permanent magnet* or an *electromagnet.*
free electron An *electron* not confined to the *electron cloud* of an *atom.* Usually an outer electron that is free to move when *electric* or *magnetic forces* act on it.
frequency Formula symbol: **F.** The number of complete *cycles* or vibrations per second performed by an *alternating current.* Units of measurement: English, *cycles per second; metric (S.I.),* the *hertz.*
fuse A *circuit protection device* whose *fuse link* is connected in *serie*s with the *circuit* to be protected.
fuse link The thin metal strip in a *fuse.* It is composed of an alloy with an extremely low melting point.
fusetron *Plug fuse* with a special time-delay *fuse link* which does not melt rapidly when a temporary *overload current* flows through it.
generator A machine or device which converts some other form of *energy* into *electricity.*
glow lamp A lamp without *filament.* Gas at low pressure is *ionized* and made to glow around the *negative* electrode.
ground A conducting path connecting a *circuit* to the earth. Metal conduits, junction boxes, switch and outlet boxes, and the metal casings of appliances must be grounded for safety.
heating effect The heat generated when *electrons* encounter *resistance* in a *conductor.* This effect is put to work in *fuses,* lamps, electric heaters, soldering irons, etc.
heating element A resistance-wire *coil* in which *electron energy* is converted to heat.
HEPC regulations Abbreviation for Hydroelectric Power Commission of Ontario regulations. A set of binding rules and regulations which govern the design and construction of *electrical circuits* and devices in the Province of Ontario.
hertz Formula symbol: **Hz.** *Metric (S.I.)* unit of *frequency.* One hertz is equal to one *cycle per second.*
incandescent lamp Electric lamp with a finely coiled tungsten *filament* which when heated to incandescence by an *electron current* gives off a brilliant light.
induced magnetism The alignment of the *magnetic atoms* of iron or steel by an external *magnetic force.* In steel, induced magnetism is *permanent*; in iron, *temporary.*
induced voltage A *voltage* generated in a *conductor* or *coil* by a changing or moving *magnetic force.*
insulator Also called *dielectric.* A material which contains few or no *free electrons* and which therefore does not conduct *electricity.*
International System of Units Abbreviated **S.I.** The system of *metric* units of measurement adopted by the 11th General (International) Conference on Weights and Measures in Paris, 1960.
interrupter contacts A mechanical device which continually makes and breaks a *circuit,* thus interrupting the *electron flow* at regular intervals.
ion An electrically charged, or unbalanced *atom* which has either a lack of

electrons (positive ion) or a surplus of electrons *(negative ion).*

ionize To change *atoms* into *ions* by either adding or removing *electrons.*

iron-filings pattern The characteristic pattern formed by iron filings when they are aligned by a *magnetic force.*

joint Connection of two or more *conductors* which must be soldered and insulated unless made with *solderless wiring connectors.*

joule Formula symbol: **J.** *Metric (S.I.)* unit of *energy* or work.

jumper wire A short length of wire used for making temporary connections in experimental *circuits.*

junction box A sheet-metal box with cover which provides mechanical protection for electrical *joints.* The *electrical safety code* requires that all joints in electrical *conductors* must be made inside junction boxes.

keeper A soft-iron bar which is kept in front of the *poles* of a horseshoe *magnet* to close the magnetic circuit while the magnet is not in use.

kilo Formula symbol: **k.** Metric (S.I.) prefix before units of measurement; it stands for a factor of 1,000 (10^3). Thus, one kilovolt is equal to 1,000 *volts.*

kilowatt-hour Formula symbol: **kWh.** *Metric unit* of *electric energy* equal to 1,000 *watt-hours* or 3,600,000 *watt-seconds* (joules).

knife-blade fuse A tubular or *cartridge fuse* with blade contacts. It is used for *current* ratings greater than 60A.

knife switch An electrical *switch* whose main part is a hinged metal blade which makes or *breaks* the circuit.

laminated core A soft-iron *core* for an *electromagnet* or a *transformer.* It is made by bonding together thin sheets of soft steel with an insulating material. The *insulator* layers reduce *eddy currents* induced in the core.

laminated magnet A *permanent magnet* made by bonding together two or more individual *magnets* in parallel, with like *poles* adjacent.

law of electric charges Like *charges* repel, unlike charges attract.

law of magnetic poles Like *poles* repel, unlike poles attract.

lead-acid cell A *secondary cell* in which lead oxides change their composition during *charging* and *discharging.* The *electrolyte* is dilute sulphuric acid.

left-hand rule For a current-carrying *conductor*: grasp the wire with the left hand, thumb pointing in the direction of *electron flow.* The fingers curled around the conductor indicate the direction of the electromagnetic force. For a current-carrying *coil:* grasp the coil with the left hand, curled fingers pointing in the direction of the *electron flow* through the coil. The thumb then points toward the *N-pole* of the electromagnetic field.

lightning Sudden, visible *discharge* of atmospheric *electricity* from one cloud to another, or from a cloud to the earth, or vice versa.

lightning arrester *Circuit protection device* which provides a *ground* path for atmospheric *electricity.*

lightning rod Grounded metal rod attached to the highest point of tall buildings to protect them against *lightning* strokes.

lines of force Alternate name: *flux lines.* The invisible paths along which *magnetic* and *electric force* acts in the space surrounding a *magnet* or an electrically *charged* body.

live circuit A *circuit* connected to a *voltage* source.

load Any electrical device which, when connected to a *circuit,* uses the *energy* carried by the *free electrons* and converts that energy into some other useful form.

lodestone Old English term for *magnetite,* or pieces of naturally *magnetic* iron ore.

loss Generally, the electric *energy* dissipated in a *conductor* or in a *load* in the form of undesired heat.

low-voltage power supply A device which converts the 115 VAC line *voltage* to a low, variable *d.c.* voltage, generally below 30 volts.

magnet An object which produces an invisible field of *magnetic force* in the surrounding space.

magnetic atom Name given to *atoms* of iron, nickel and cobalt, all of which are tiny, ball-shaped *permanent magnets.* Alternate names: magnetic dipoles and atomic dipoles.

magnetic compass An instrument for indicating geographic directions by means of a pivoted, permanently magnetic needle whose *N-pole* always points toward the earth's *magnetic north pole* in Northern Canada.

magnetic energy *Magnetic force* doing work, such as inducing an *electron current* in a *conductor.*

magnetic field The region or three-dimensional zone around a *magnet* or *electromagnet* in which the total magnetic *flux* of the device can be detected.

magnetic force The invisible force of attraction (or repulsion) which a *magnet* exerts on *magnetic materials.*

magnetic materials Iron, nickel, cobalt and their alloys.

magnetic north The direction indicated by the *N-pole* of a suspended *magnet* or a *compass needle.* Does not coincide with true geographic north because the two poles are approximately 1,100 miles apart.

magnetic pole One of two points on any *permanent magnet* or *electromagnet* at which the intensity of the *magnetic force* is strongest.

magnetic saturation The condition in a *magnetic material* when all its *magnetic atoms* are aligned in the same direction, or when it contains all the *flux lines* which it can hold.

magnetic shield Soft-iron casing for protecting delicate instruments from the influence of *lines of* magnetic *force.*

magnetism The invisible force acting in the space surrounding a *magnet.*

magnetite Scientific name for naturally magnetic iron ore.

magnetizing force The amount of *magnetomotive force* acting on each inch (or metre) of a *magnetic material.* Units of measurement: English, *ampere-turns* per inch; Metric, ampere-turns per metre.

magnetomotive force The force which produces *flux lines* in a *magnetic material* or in space. Unit of measurement: the *ampere-turn.*

magnet wire Thin copper wire, usually insulated with an especially flexible coating of vinyl or other high-quality insulating material.

mega Formula symbol: **M.** *Metric (S.I.)* prefix for units of measurement. Mega stands for a factor of 1,000,000 (10^6). Thus, one megavolt (1 MV) is equal to 1,000,000 *volts.*

metric prefixes A system of special words which, when added in front of the name of a basic *metric* unit, converts that unit to a decimal multiple or a decimal fraction of its basic value.

metric system A coherent system of units of measurement based on the metre, kilogram and second as its basic, arbitrarily defined units. All other units, including the electrical ones, are defined in terms of these basic units. In 1960, the 11th General (International) Conference on Weights and Measures adopted the metric system as the only international system of measurements and named it the *International System of Units,* abbreviated *S.I.*

micro Formula symbol: μ. *Metric (S.I.)* prefix for units of measurement. Micro stands for a factor of 1/1,000,000 (10^{-6}). Thus, 1 microampere is equal to 1/1,000,000 of an *ampere.*

mil-foot English unit of measurement for *conductor* volume, equal to a round wire of one-foot length with a diameter of one mil, or 1/1,000 of an inch. The mil-foot is used in the measurement of the *specific resistance,* or *resistivity,* of a conductor material.

milli Formula symbol: **m.** *Metric (S.I.)* prefix for units of measurement, equal to a factor of 1/1,000 (10^{-3}). Thus, one milliampere is equal to 1/1,000 of an *ampere.*

mogul base Large screw base for *incandescent lamps* having high power ratings.

natural magnet A piece of *magnetite* ore which has a weak *magnetic field.* Also called *lodestone*.

negative Having an excess or surplus of *electrons.*

negative charge The name given to the natural *electric charge* of *electrons* which enables them to exert an invisible *electric force* in the surrounding space.

negative ion An electrically unbalanced *atom* that has an excess of *electrons* in its *electron cloud.* Also, an atom with a *negative* electric *charge.*

negative terminal That *terminal* of a *voltage* source or *power supply* which has an excess of *electrons,* and which therefore exerts a pushing force on the *free electrons* in a circuit connected to the source.

neutral Possessing neither a *negative* nor a *positive electric charge.* A neutral body consists of a like number of *protons* and *electrons* and therefore has no net *electric charge.*

neutron Heaviest of the three *elementary particles;* it has no *electric charge.*

nickel-cadmium cell Storage *cell* having a cadmium *(negative)* electrode and a nickel *(positive)* electrode immersed in a paste *electrolyte.*

non-magnetic materials All substances which are not affected by *magnetic force.*

normally closed Term used to identify *relay* or *switch contacts* which are closed in their normal position, and which *break* the *circuit* when the *switch* is operated.

normally open Term used to identify *relay* or *switch contacts* which are open in their normal position, and which close the *circuit* when the *switch* is actuated.

N-pole Abbreviation for northseeking *pole*; that pole of a *permanent magnet* or *electromagnet* which, when free to turn, points toward the *north magnetic pole* of the earth.

north magnetic pole The magnetic north pole of the earth is located some 1,100 miles (1,800 km) south of the geographic north pole.

nucleus The central part or core of an *atom,* composed of *protons* and *neutrons.*

ohm Formula symbol: Ω . *Metric (S.I.)* unit of electrical *resistance.* The resistance of a *circuit* in which an *emf* of one *volt* maintains a *current* of one *ampere.*

Ohm's law A law stating the relationship between *voltage, current* and *resistance* in an electric *circuit.*

ohmmeter An instrument for the measurement of electrical *resistance.*

Ontario Electrical Code Alternate name for the *HEPC Code* of Ontario.

open circuit A *circuit* containing a break in the metallic pathway for *electrons.*

OR gate An arrangement of two or more *switches* in parallel, each of which, or all, can close the *circuit.*

output voltage The *terminal voltage* of a *cell, battery, power supply,* or other source of *electricity.*

overload A *load* in excess of that for which a *circuit* was designed.

parallel circuit A *circuit* in which all *loads* receive the same *voltage* from one source, and in which the total *current* divides into the individual load currents. A circuit in which each load can operate independently from all other loads.

permanent magnet A piece of hard steel or other alloy that has been strongly magnetized and will retain its *magnetism* indefinitely.

permeability The ability of a material to conduct *lines of magnetic force*.

photovoltaic cell A *solar cell* which converts light *energy* to a low-voltage *direct current.*

pictorial diagram A *circuit* diagram in which drawings of the components and wires are used in place of *schematic symbols.*

piezoelectric crystal A crystal having the unique property of generating *electricity* when a changing pressure is applied to its surfaces.

pitting Erosion of the surface of electrical *contacts* caused by excessive arcing when the contacts are opened under *load.*

plug cap Removable connector attached to the end of a linecord or other *conductor* for insertion into a *receptacle.*

plug fuse A screw-base *fuse* for use in standard home fuse panels, with a *voltage* rating of 125 *volts* and a maximum *current* rating of 30 *amperes.*

plunger The movable, spring-loaded magnetic core of a *solenoid.*

polarity The characteristic of having opposite *magnetic* or electric *poles.*

positive Having a lack or deficiency of *electrons.*

positive charge The name given to the natural *electric charge* of *protons* which enables them to exert an invisible *electric force* in the surrounding space.

positive ion An electrically unbalanced *atom* which has a deficiency of *electrons* in its *electron cloud.*

positive terminal That *terminal* of a *voltage* source or *power supply* which has a lack of *free electrons,* and which therefore exerts a pulling force on the free electrons in the *conductor* connected to it.

potential energy The *energy* stored in stationary *electric charges.*

power Formula symbol: **P.** The rate at which *energy* is used in an *electric circuit* or in a *load*. Unit of mesurement: the watt.

power supply That part of an *electric circuit* which gives energy to *free electrons.* Usually a *battery, transformer, generator* or other *energy converter.* Also called source, voltage source, or supply.

primary cell A device which converts chemical energy to *electrical energy* by using up its ingredients, and which therefore cannot be recharged. Also called a *voltaic cell.*

primary coil The input *coil* of a *transformer* in which the incoming *alternating current* is converted to a moving *magnetic force.*

proton *Elementary particle* of much greater mass than an *electron.* Carries a permanent *electric charge* equal to and opposite to that on the electron.

push-button switch A spring-loaded *switch* designed to be operated by finger pressure.

quartz-iodine lamp A small and lightweight tungsten *incandescent lamp* filled with iodine gas and having a quartz envelope.

rat-tail joint Electrical *joint* of two or more wires twisted together.

receptacle Electrical outlet; usually two U-ground outlets in a common frame, one on top of the other.

relative resistance The *resistance* of a metal compared to that of a copper *conductor* of equal size.

relay An electromagnetically operated *switch* which can be opened or closed from a remote location.

reluctance The opposition of a *magnetic material* to the setting up of *flux lines.*

residual magnetism *Magnetism* remaining in a *magnetic material* after the *magnetizing force* has been removed.

resistance Formula symbol: **R.** The opposition which *electrons* encounter when flowing through a *conductor.* Unit of measurement: the *ohm.*

resistivity The *resistance* of a unit volume of a material. In the English system, the resistance in *ohms* per one *mil-foot* length of a *conductor.* In the *metric (S.I.)* system, the resistance in ohms per one metre cube of a material. Alternate name: *specific resistance.*

resistor An electrical device having a definite amount of *resistance* concentrated in a small space.

retentivity The ability of a *magnetic material* to retain *magnetism* after the *magnetizing force* has been removed.

return wire Name given to that *conductor* in an *electric circuit* which connects the *load* directly to the *power supply.*

ring magnet A ring-shaped *permanent magnet* whose *magnetic force* is contained entirely within the magnet. Also, a ring-shaped flat magnet whose *poles* are distributed along the opposite surfaces of the magnetic disc.

rosin Non-corrosive *flux* for electrical and electronic soldering. Also called colophony.
rosin joint A defective soldered *joint* which was insufficiently heated during soldering and which makes only intermittent contact.
safety first The golden rule of working with electricity: safe habits save trouble.
schematic diagram A *circuit diagram* in which the *conductors* and circuit *components* are represented by graphic symbols.
schematic symbol A graphic symbol representing an electrical *component* on a *schematic diagram.*
secondary cell A rechargeable storage *cell,* the basic unit of a *storage battery*.
secondary coil The output *coil* of a *transformer.*
semiconductor A material which conducts *electricity* only under special conditions.
series circuit An *electric circuit* in which all *components* are connected in single file, or end to end, to form a single path for *free electrons.*
series-parallel circuit A circuit which combines *series* and *parallel circuits* in any combination into a single *electric circuit.*
set-screw connector A *solderless wiring connector* consisting of a copper ferrule with a set screw and a detachable insulating jacket.
shelf life The time period for which an unused *cell* or *battery* can be stored before losing its ability to generate *electricity.*
short circuit An accidental low-resistance *current path* that bypasses the *load* of an *electric circuit.*
S.I. Abbreviation for *International System of Units.*
signal The message or information in the form of a small signal *current* through an electrical communications system.
signal device Any electrical device which converts an electrical *signal* into an audible or visible signal, such as a bell, *buzzer* or *annunciator.*
silicon A pure *semiconductor* material used in the manufacture of *solar cells* and transistors.
solar cell Also called *photovoltaic cell.* A device which converts light *energy* to *electricity.*
solder A low-melting-point alloy of lead and tin which is used to bond together copper *conductors* in electrical and electronic *circuits*.
solder bridge Small bridge of liquid *solder* which transfers heat energy from the soldering iron into the work to be soldered. It also promotes the spreading of the *rosin flux.*
solderless wiring connector Special *twist-on, set-screw* or *crimp-on connector* for making electrical *joints* without the need for soldering and extra insulation.
solenoid *Electromagnet* with a movable *core* or *plunger* that can be used to actuate *switches,* open door locks, strike chimes, and perform other mechanical operations.
source Alternate name for *power supply* or *voltage* source.
south magnetic pole The *magnetic* south pole of the earth, located approximately 1,500 miles (2,500 km) from the geographic south pole in the antarctic region.
specific resistance The *resistance* of a one *mil-foot* length of a *conductor,* measured in *ohms* per mil-foot. Its *metric (S.I.)* equivalent is the ohm/metre.
splicing compound A rubber-base insulating tape which is used for insulating high-voltage circuit splices.
S-pole The southseeking *pole* of a *permanent magnet* or *electromagnet.* The pole opposite to the *N-pole.*
SPST Abbreviation for single-pole, single throw; an electrical on-off *switch*.
static electricity A stationary *electric charge* on a body.
storage battery A *battery* composed of storage or *secondary cells,* chiefly *of lead-acid* or *nickel-cadmium cells.*
supply Short form for *power supply,* that part of an *electric circuit* which gives *energy* to the *free electrons.*
switch A device for making or breaking

the metallic pathway for *free electrons* in an *electric circuit*.

tap joint Alternate name: tee *joint*. An electrical connection of a branch *conductor* to a long-run *conductor*.

temporary magnetism *Magnetism* induced in soft iron by an outside *magnetic force*. It disappears when the external magnetic force is removed.

terminal A screw, *binding post*, soldering lug or other point to which electrical connections can be made. More generally, the two *poles* of a *battery* or *power supply* to which a *circuit* is connected.

thermocouple A welded junction of two unlike metals which when heated generates a flow or *electron current*, thus converting heat energy to *electricity*.

thermopile Several *thermocouples* connected in *series* around a flame or other source of high temperature to increase the available output *voltage*.

transducer A device which changes electrical *signals* to sound or mechanical vibrations, or vice versa.

transformer A device which changes *alternating current* to different *voltage* and *current* levels by *electromagnetic induction*.

twist-on connector A *solderless wiring connector* with an integral insulating jacket which is twisted on the *conductors* to be joined.

U-ground plug Special three-prong *plug cap* with a U-shaped grounding prong. The three prongs are colour-coded to identify the *conductors* to be connected to them: brass colour, black *feeder wire;* silver colour, white *return wire;* green prong, green *ground* wire.

U-ground receptacle Special receptacle for *U-ground plugs* which is installed in all modern house-wiring *circuits* to provide grounding for appliances.

Underwriter's Laboratories, Inc. Abbreviated **UL.** American testing and approving agency for electrical equipment. Similar to the *Canadian Standards Association,* or *CSA.*

volt Formula symbol: **V.** *Metric (S.I.)* unit of measurement for *electromotive force* or *voltage.*

voltage Formula symbol: **E.** Alternate name for *electromotive force* or potential difference.

voltaic cell The original *primary cell* invented by Alessandro Volta, consisting of two unlike metals immersed in an acid solution, or *electrolyte.*

voltmeter Instrument for measuring the *voltage* of a *power supply* or other voltage source.

watt Formula symbol: **W.** *Metric (S.I.)* unit of measurement for *power.* One watt is equal to a rate of *energy* use of one *joule* per second.

watt-hour Formula symbol: **Wh.** Unit of measurement for *energy;* equal to 3,600 *joules.*

wattmeter An instrument for measuring the *power* consumption of an *electric circuit* or device. A wattmeter measures *current* and *voltage* simultaneously and automatically computes and indicates the product of the two as *power* in *watts.*

watt-second Formula symbol: **Ws.** Unit of measurement for *energy,* equal to one *joule,* or a *power* of one *watt* delivered for a period of one second.

western union joint A special splice used for joining long-run *conductors;* it was invented by linemen of the Western Union Telegraph Company in the nineteenth century.